I0502947

Construction Methods
For Civil Engineers

About Book   =

In this book, the question and its answers are given in full practical form. This book is solely for the construction industry. In this, we have put the full question and its answer from our level .Whether a planning manager or project manager should have the level of question and its answer, almost all of them are included in this book .This book has been written keeping the interview in mind .

OUR TEAM – OUR STRENGTH
 ➢ Our Professional team consists of :
 ➢ Project Managers
 ➢ Civil Engineers
 ➢ Interior Designers
 ➢ Planning Engineers
 ➢ Billing Engineers
 ➢ MEP Engineers
 ➢ Quantity Surveyors
 ➢ Quality Engineers
 ➢ Safety Engineers
 ➢ Project coordinators
 ➢ A brief profile of Leadership Team is given ahead

Our Services
 ➢ Construction Project Management consulting
 ➢ Contract & Cost Management
 ➢ Quantity Survey ,Estimation & Tender Preparation ,Architecture and Structural design
 ➢ BBS Preparation
 ➢ Billing & Cost Management
 ➢ Third Party audit / and evaluation of RA and Final Bills.
 ➢ Project management Advisory
 ➢ Quality audit
 ➢ Safety audit
 ➢ Land Surveying
 ➢ Landscaping
 ➢ Vertical Gardening
 ➢ Vendor Allotment for All types of Construction activities

About Author & About my Team –

Nitin Gupta is the Managing Director & Principal Consultant of RN Infra Projects, which provides Project Management Services, Cost Management and Construction Management Consulting Services to various Clients in Residential, Commercial, Industrial and manufacturing IT Parks, Infrastructure, Marine, Education, Leisure, Health and Hospitality sectors. Nitin Gupta is a Civil Engineering graduate with over 8 years of experience in various sectors in which 2 years in Factory /Cold store Contract Administration and 6 years in Project Management, Construction & Cost Management Services.

➢ He has a vast experience in the field of infrastructure and constructions .Starting from an early age he also has diverse experiences in the area of Mining and Trading .He has visited an innumerable number of mines in India .He is an etiquette Man, focusing on pursuing global network . With self-taught study and research, he has impeccable ideas to grow, to flourish and to attain a reputed name for his company.
➢ With twinkling photons, his ambition is to further contribute reaching the pinnacle of global Standers.

2- Assistant Director : **Mr. Ashish Jain**

Educational Qualification : M.Tech In structural Engineering
Total experience : 7 Years
Positions held : Site Manager
Sr. Project Engineer at Godrej, Bangalore Project.

Responsibilities handled - :

> *Team Leader - Site Execution of Laboratory Projects at DRL, Hyderabad.*
> *Planning, Scheduling site activities & labour resources.*
> *Site Coordination & review of various team's work.*
> *QA and cost control – review and reporting.*
> *Contractor bill certification; Project Closure Documentation.*

3- Executive Director (North Zone) : Mr.Dharamraj

Qualification : Civil Engineer.
Total experience : 15+ years.
Positions held : Pacific Infra Projects
Since Oct.15 Reinforce Construction Pvt. Ltd., New Delhi ,PS Infracon (Bangalore)

Responsibilities handled:

> *Presently Project-In-Charge for Cold Storage Project at Tumkur.*
> *Assistant Project Manager for Industrial Project at Dr. reddy laboratory.*
> *Tendering, Vendor Evaluation & Material Sourcing.*
> *Estimation, Tender submission & Rate Analysis.*
> *Construction Management – Scheduling and Billing.*
> *Certification of JMS, Rata Analysis, Billing and lessoning with various Govt. Agencies.*

Project in Charge

Sub :   Submission of Methodology of Tile flooring work.

Dear Sir,

Please find enclosed the copy of methodology for carrying out Tile flooring work at Project Name.

For your record please.

For Company Name

Methodology of Tile flooring Work

CLIENT- Name

Revision-00

	Prepared By	Review By	Approved By
Designatio n			
Signature			
Name			
Date			

Methodology for Indian Tile Flooring Work

Objective

Objective of this procedure is to provide a guideline for sequence of operations pertaining to IPS flooring work, so as to ensure that the works are carried out in a systematic manner and to ensure that the works conform to the drawings and technical specification.

Scope

This procedure is applicant for Tile Flooring work as per approved drawing.

Reference

This procedure is based on the requirement of the approved drawings, technical specification and relevant IS Codes.

- **BOQ.**

1.8	Tile Flooring.	UNIT
	laying to pattern homogeneous tile flooring of approved size, , make and shade and free from cracks, warpage, fissures and flakes with straight edges and perfect corner laid at right angle as per design over a bedding of cement mortar 1:4 (1 cement : 4 coarse sand) of adequate thickness to match the total thickness of 50mm, finishing the joints neatly with epoxy grout approved by GPL of Laticrete make or equivalent mixed with matching shade, including cutting, epoxy grouting the joints and protection etc. complete in all respects as per pattern and drawing. 02 mm PVC spacer shall be used of Arpitha make or equivalent as approved by Project Manager.	Sqm

Resource Deployment.

- The Contractor shall deploy any or all of the following resources as dictated by the volume and nature of work.

Material

- Approved Cement.

- Specified Tiles.

- Portable water.

- Coarse Sand.

- 2mm Approved Spacer

Tools & Tackles

- Chisel

- Hammer

- Bucket

- Safety PPE

- Aluminum Straight edge.

- Spirit Level and other leveling Equipment.

- Trowels.

- Nylon thread.

Preparatory Works.

- Construction surface area should be chipped and cleaned prior to start of work. Any loose material, dust shall be cleaned with wire brush or as required.
- Work surface of contaminated with oil or grease, shall be cleaned thoroughly to remove such contaminants.

Application of Works

- Applying a coat of cement slurry for proper holding of cement mixture.
- Tile cement mortar shall be 1:4 as specified.
- The average thickness of cement mortar shall be 50mm including tile thickness.
- Motor shall be spread, tamped & corrected to proper level & slopes.
- Over cement mortar neat grey cement slurry shall be spread over for fine bonding with tile.
- Tiles shall be cleaned and fixed with spacer 2mm between tiles.
- Each tile being gently tapped till it is properly in line & level.

Finish

- After completion of work check level maintain proper level on the surface & tile need to be cleaned
- All the tile groove need to fill with approved cement based grout.
- No traffic shall be allowed on bedded tiles at least for two days.
- Carrying out curing for minimum of 3 days.
- For all trowel finished surfaces, the finishing has to be completed before the mortar of concrete is fully set.

Tolerance

- Tolerance in the final finish could be 3 to 5mm (+) (-).

Quality Test.

Cement:

- Contractor shall obtained test certificates on the physical & chemical properties of cement from the manufactures for each batch of cement received to ensure conformity with relevant IS specification and keep the certificate on record with a copy to GPL.

- In addition with a view to cross check the manufactures test result sample from first consignment of a particular brand shall be test from GPL approved lab periodic validation of the manufacture test result shall be carried out by testing the physical & chemical properties of cement in GPL approved lab at every six month per source .

Fine Aggregate (Sand):

- Testing of sand which is used in Tile flooring shall be as per Inspection Test Plan (ITP)of GPL.

Water:

- Testing of water which is used in Tile flooring shall be as per Inspection Test Plan (ITP) of GPL.

<u>Safety Consideration and Requirement.</u>

- Each workman should use all PPEs like hand gloves, nose musk, safety goggles, safety belt, helmet and safety shoes during Particular works.
- All work will be instructed as per GPL work inspection.

<u>Date</u>

Project in Charge

Sub   :   Submission of Methodology of Sunken Filling Work

Dear Sir,

Please find enclosed the copy of methodology for carrying out Sunken Filling stonework (IPS) work at Godrej Nature + project.

For your record please.

<u>For Company Name .</u>

Methodology of Sunken Filling Work.

CLIENT- GODREJ
HIGHVIEW LLP

Revision-00

	Prepared By	Review By	Approved By
Designation			
Signature			
Name			
Date			

Methodology of Sunken Filling Work.

Objective.

- Objective of this procedure is to provide a guideline for sequence of operations pertaining to Sunken Filling work, So as to ensure that the works are carried out in a systematic manner and to ensure that the works conform to the drawings and technical specification.

Scope.

- This procedure is applicable for Sunken Filling work in structure.

REFERENCE.

- This Procedure is based on the requirements of the GFC Drawings and Technical Specification.

BOQ.

3	Sunken Area Filling	UNIT
	Providing and laying on sunken areas over waterproofing courses, CC screed with 1:4:8 (1 Cement: 4 coarse sand: 8 graded stone aggregate 20 mm nominal size) concrete in required slope etc. complete. (Screeding shall be done after laying of plumbing pipes)	

Resource Deployment.

* The Contractor shall deploy any or all of the following resources as dictated by the volume and nature of work.

Material

* Approved Cement.

* Portable water.

* Coarse Sand.

* P.P.E. As required.

* Aggregate.

Tools & Tackles

- Chisel

- Hammer

- Bucket

- Brush

- Mixing Tools

- Safety PPE (Safety shoes, Helmet, Rubber Hand Gloves, Goggles, Nose Mask Etc.)

- Aluminum Straight edge.

- Spirit Level and other leveling Equipment.

- Trowels.

- Nylon thread.

- Mixture Machine.

- Spade/Mortar hoe.

Preparatory Works.

- Construction joints surface area should be chipped and cleaned prior to start of work. Any loose material, dust shall be cleaned with wire brush or as required.
- Work surface of contaminated with oil or grease, shall be cleaned thoroughly to remove such contaminants.

Application of Works

- Applying a coat of cement slurry for proper holding of cement mixture.

- Glass/PVC strip fixing should be done with cement mortar (1 cement: 4 Sand) at least one day in advance.

- Do the concreting in the continues sequence in the ratio of with 1:4:8 (1 Cement: 4 coarse sand: 8 graded stone aggregate.

Finish

- After completion of concrete check level maintain proper level on the surface & divider strips top should be visible.

- Carrying out curing for specified period of.

Tolerance

- Tolerance in the final finish could be 5 to 10mm (+) (-).

Quality Test.

Cement:

- Contractor shall obtained test certificates on the physical & chemical properties of cement from the manufactures for each batch of cement received to ensure conformity with relevant IS specification and keep the certificate on record with a copy to GPL.

- In addition with a view to cross check the manufactures test result sample from first consignment of a particular brand shall be test from GPL approved lab periodic validation of the manufacture test result shall be carried out by testing the physical & chemical properties of cement in GPL approved lab at every six month per source .

Fine Aggregate (Sand):

- Testing of sand, which is used in IPS flooring, shall be as per Inspection Test Plan (ITP) of GPL.

Aggregate:

- Testing of aggregate, which is used in IPS flooring, shall be as per Inspection Test Plan (ITP) of GPL.

Water:

- Testing of water, which is used in IPS flooring, shall be as per Inspection Test Plan (ITP) of GPL.

Safety Consideration and Requirement.

- Each workman should use all PPEs like hand gloves, nose musk, safety goggles, safety belt, helmet and safety shoes during particular works.
- All work will be instructed as per GPL work inspection.

Methodology of Gypsum Boxing Work.

Project	Revision
Godrej Nature +	**R00**

	Prepared By	Checked By	Review By	Approved By
Designation				
Signature				
Name				
Date				

Methodology of Gypsum Boxing Work.

Objective:

- Method statement and procedure for gypsum boxing work.

Scope:

- This procedure is applicable gypsum boxing working structure.

Reference:

- This Procedure is based on GFC Drawings and Technical Specification BOQ.
- IS- 2095 (1964, 1982) Pt-1, 3 for False Ceiling.
- IS -2441-1984 for Nails and Screws used in Gypsum Boxing.
- ASTM C475 / EN 13963 for Joint Tapes & Compounds.

BOQ:

Providing and fixing **suspended false ceiling** which includes providing and fixing GI perimeter channels of size 0.55 mm. Thk. having one flange of 20 mm and another flange of 30 mm and web of 27 mm along the perimeter of the ceiling, screw fixed to brick wall/ partition with help of nylon sleeves and screws, at 610 mm centers. The suspended GI intermediate channels of size 45 mm,0.9mm thk with two flanges of 15 mm each from the soffit at 1220 mm centers with ceiling angles with ceiling angle of width 25mm x 10mm x 0.55mm thk fixed to soffit with GI cleat and steel expansion fasteners. Ceiling section of 0.55 mm thk having web of 51.5 mm and two flanges of 26 mm each with lips of 10.5mm are then fixed to the intermediate channels with help of connecting clips and in direction perpendicular to the intermediate channel with centers. All GI members shall be of Gypsteel or approved equivalent make. The 12.5 mm tapered edge Gypboard (Confirming to IS-2095-1982) is then screw fixed to ceiling section with 25 mm dry wall screws at 230mm centers.

Resource Deployment.
- The Contractor shall deploy any or all of the following resources as dictated by the volume and nature of work.

Material
- Approved GI perimeter channels of size 0.55 mmThk. having one flange of 20 mm and another flange of 30 mm and web of 27 mm **(PERIMETER)**Gyproc.
- Approved suspended GI intermediate channels of size 45 mm, 0.9mm thick with two flanges of 15 mm.
- Ceiling angle of width 25mm x 10mm x 0.55mm **(MAIN CHANNEL)**Gyproc.
- G I soffit Cleat 37 X27X25mm thickness 1.6mm Gyproc.

- Ceiling section of 0.50 mm thk having web of 51.5 mm and two flanges of 26 mm each with lips of 10.5mm **(CEILING SECTION)** Gyproc.
- Steel expansion fasteners Gyprock 40X8mm.
- 12.5 mm tapered edge gypboard.
- 50mm width JointPaper Tape Gyproc.
- Pro-Fill jointing compound Gyproc.
- 25mm drywall screw Gyproc.

Tools & Tackles
- Drill.
- Hammer Drill.
- Screw Driver.
- Big Scissor.
- Empty Bucket for Compound
- Gurmala &Patti.
- Other small tools as per requirement for convenience
- Safety PPE (Safety shoes, Helmet, Rubber Hand Gloves, Goggles, Etc.)

Surface Preparation:
- Ensure the line / level / plumb of the wall and if any major deviation observed the same shall be rectified.
- Ensure the materials received checked for approved brand of the manufacturer and original packing before accepting the same. (Proposed Gypsum board brands- Gyproc / Equivalent)
- Ensure the wall surface/ ceiling slab surface neatly cleaned for loose materials or dust particles
- Ensure The pipe works to be concealed properly checked for any leakage
- Ensure the MEP services completed including the hole filling and approval taken from the concerned department

- Ensure the boxing layout checked for the positions and approval taken from the engineer.
- Proper safe and stable working platforms should be available for boxing works in the ceiling
- Ensure that all works of Fire Fighting, Electrical, and Gypsum Plaster on walls and ceiling should be completed before starting boxing works.

Application of works:
- Marking level for bottom of boxing +12.5mm on wall with colour thread to fix perimeter as per drawing.
- Fixing perimeter on wall with fastener at 1220 mm C/C.
- Fixing of soffit cleat to ceiling 1220mm C/C with fastener.
- Fixing of Ceiling angle with cleat.
- Fixing main channel to be hanged with angle at 1220mmC/C in ceiling.
- Fixing perimeter on ceiling with fastener at 1220 mm C/C.
- Fixing ceiling section on bottom &vertical outer face at 1220 C/C.
- Fixing gypsum board at bottom to be screwed at 230 mm C/C with main & perimeter on wall.
- Fixing perimeter at bottom screwed with board at 230mm c/c.
- Fixing of Gypsum board vertical screwed with perimeter in ceiling & beam bottom.

Post Check of Gypsum Boxing Activity:
- The line, level and plumb of the completed works has to be checked using spirit level
- The gypsum board have to checked for any damaged edges, cracks after fixing
- All the drywall screws should be fully driven in, dose not project out of the board

- The angles, channels have to be checked for any bend or damage during fixing
- The minor gap of 2-3mm observed between the wall and boxing works shall be sealed with silicon sealant.
- Rectifications of post observations if any, the final finishing activities can be carried out

Material Testing Guidelines:

- Testing of materials, which is used in gypsum boxing work, shall be as per GPL Inspection Test Plan.

Material Storage Guideline:

- Ensure the GI Frame work material for false ceiling works not stored in the open and exposed to direct sun and rain. Suitable covered storage space provided for the safe storage. Check height limitations when stacking materials, it should be from 1.5 to 2.0 m height.

- Limit stacks of gypsum boards to a maximum height of 10 units or less based on conditions

- The gypsum boards should be stored on wooden battens on a flat surface and have to be stacked in piles of smaller heights and not to be stacked on its edges, as the edges gets damaged

- Store gypsum boards in a cool dry place, and keep away from direct sunlight. Keep in Frost-free and best before use 12 months /as per manufacturer recommendation

- The nails & Screws to be stacked in bagged by stepping back the layers and cross keying the bags at least every 10 bags high, except when restrained by walls or partitions of adequate strength.

Safety Consideration and Requirement:
- Each worker should use all PPEs as hand gloves, nose mask, safety goggles, and safety belt, helmet and safety shoes during the works.
- All work will be instructed as per GPL work inspection.

Methodology of Floor Tile Work.

Methodology of Floor Tiling Work.

Objective:

Objective of this procedure is to provide a guideline for sequence of operations pertaining to floor tiling work, so as to ensure that the works are carried out in a systematic manner and to ensure that the works conform to the drawings, BOQ, SOP and technical specification.

Scope:

This procedure is applicant for Tile Flooring work as per approved drawing methodology and SOP.

Reference:

This procedure is based on the requirement of BOQ, approved drawings, technical specification, SOP and relevant IS Codes.

Resource Deployment:

The Contractor shall deploy of the following resources as dictated by the volume and nature of work.

Tools & Tackles:
- Trowel
- Nylon thread
- Spirit level
- Aluminum straight edge
- Rubber Mallet

- Right Angle
- Tile Cutting Machine
- Spade
- Mortar Pan
- Half round tool
- P.P.E. as required

Materials:
- Tiles (Vitrified, Antiskid, ceramic, Glazed)
- Coarse Sand
- Potable Water
- Approved Cement of approved make
- 2mm thick spacer (up to 600* 600)
- Masking Tape
- Protection Sheet

Procedure:

SURFACE PREPARATION & JOINTING / BEDDING MORTAR:
1) Check availability of latest architectural drawing on site.
2) Complete Electrical & plumbing work before starting of tile work.
3) MEP Clearance to be taken prior to start of flooring work
4) Ensure that quality of tile as per specification and approved sample.
5) Decide which tiles to be cut. Cut tiles to be at edges so that they are hidden inside the skirting.
6) Ensure FFL level is transferred from master level from lift lobby for all floors and same to be

transferred inside flats area where flooring need to be processed.

7) Make (RL +1 m level) from SSL with level tube. In entire flat.

8) Ensure the slab rough by tamping with hammer & clean slab with wire brush, water and broom.

9) If there is any cracks on the surface those are to be repaired with approved repairing SOP

10) Apply Anti Capillary coat with approved waterproofing material (Master seal 550 or Equivalent) to the Block work surface and to the floor.

11) Lay cement slurry 2.5Kg/m2 (1:1) so that micro cracks are filled up

12) Create level pads to check thickness of bedding mortar.

13) Vitrified tiles does not require pre-soaking, however for ceramic tiles presoaking for 2 hours is required.

14) For flooring with slopes (Toilet & Balcony / deck) following slope to be provided (to be ensured through difference in level pads)

1. Toilet: 10 mm in dry area. Further 10 mm in wet area. (1:120)
2. Balcony: 1:100.

15) If thickness is more than 40 mm then lay screed with Minimum M10 grade to be used for additional thickness.

16) Ensure electrical and plumbing openings are closed and well protected

17) Mix mortar PPC or OPC 43 grade Cement (1): Sand (5) / as per BOQ.

LAYING OF TILES:

1) Cure flooring and skirting for a minimum of 14 days.
2) Start fixing of tiles from starting point as mention in approved GFC drawing.
3) Ensure 5 mm gap to be left between wall and tiles for expansion.
4) Place the tile on bedding mortar and Follow arrow mark in tile
5) Use rubber mallet for tamping on top of tile to remove hollowness.
6) Check joints and level of tile with spirit level
7) Mix cement slurry (1:1) cement consumption 3.5Kg/m2
8) Rake the cement slurry for proper penetration.
9) Remove the spread cement slurry from the tiles.
10) Check hollowness beneath the tile by temping and checking the sound produced
11) Spacers provided while fixing floor tiles needs to be removed after setting of cement or within 48Hrs
12) After spacer removal, clean up the space between the joints.
13) Apply masking tape on all edges of tile and then Rake the joints of tiles using half round tools to fill epoxy grout of matching shade.
14) Clean floor and do not allow anyone to enter for at least 24 hours.
15) After that cover all joints of tiles with masking tape to restrain the entry of dirt and dust.
16) Now cover the whole floor are using protection sheet.

LAYING OF SKIRTING:

17) Wet the area where skirting is to be applied
18) Fix the skirting with cement paste or tile adhesive without voids behind
19) Check the line, level and right angle for skirting
20) Ensure joint of skirting matches with joint of flooring

JOINT FILLING FOR TILES WITH SPACERS

1) Clean the joint with mild detergent and sponge.
2) Remove the PVC spacers and dust with air blower.
3) For dry areas such as living room/ bedrooms , passage: use polymer modified cementitious grout
4) For wet & exterior areas such as toilets, balconies, terraces use epoxy grout
5) Apply 1 inch paper masking tape on either side of tile joint.
6) Mix the grout with water as per manufacturer's guidelines
7) The grout to be approx. 0.5 mm lower than tile top and finished in curve.
8) The excessive grout from the tile joints should be thoroughly wiped off.
9) Wait for 2 to 3 days and tamp with rubber mallet for hollowness
10) For a given tile, if hollowness observed in more than 2 out of 5 points (4 corners and center) then it shall be replaced.

MATERIAL TESTING GUIDELINES:

1) Flooring Tile for various grades of Ceramic and vitrified tile are to be tested for One Third Party Test at the start of the project and at Change of source and for every 54000SQFT/ 5000M2.

2) Field test for Every lot / Batch received for visual inspection, MTC verifications
3) Tile Grout – Third party test certificate (should not be older than 6 months from date of approval) to be submitted by manufacturer for brand approval and MTC verification on each lot received.

MATERIAL STORAGE GUIDELINES:

1) Store tiles in a dry and well-ventilated area.
2) Maintain register for each batch of tile being delivered in a register. Keep aside min 3% of each batch for purpose of attic stock, consider 3% for handling wastage, and label the attic stock properly with traceability to the laid location and batch Number.
3) Tiles shall be stacked vertically and not in a manner, that height of the stack shall not be more than 1m.

SAFETY MEASURE:

Proper precaution should be taken during tile fixing work i.e. workers should wear PPE like hand gloves, goggles, safety shoe, helmet, while working. Proper staging shall be provided where required.

Methodology of Stone Counter work

Methodology of Counter Top

Objective:

- Objective of this procedure is to provide a guideline for sequence of operations pertaining to Counter Top work, So as to ensure that the works are carried out in a systematic manner and to ensure that the work conform to the drawings and technical specification.

Scope:

- This procedure is applicable for Counter Top work in Kitchen & Toilets.

REFERENCE:

- This procedure is based on the requirement of the GFC Drawing, Technical Specification and BOQ.
- IS 14223 for granite stone

BOQ:

Providing and fixing machine cut pre-polished stone slab in required size of approved quality in counters fixed over 19mm thick water-proof ply supported with MS angle frame with hilti anchor fasteners including providing and applying two or more coats of approved synthetic enamel paint over a coat of primer including surface preparation, and including front facia, verticals, pointing with tinted white cement, polishing, curing etc. complete all as per drawings and as per detailed specifications, making openings/holes for sink, mixer in the Granite counter and joints neatly finished with white cement paste mixed with matching pigment to match the shade of stone including edge chamfering, moulding and mirror polishing and filling silicon at junctions with sanitary fixtures and corners with wall wherever required etc. complete as per drawing. Rate to include edge polishing as approved. Rate to include cost of necessary MS framework. Plan area shall be measured, with all lead and lift and as directed by Project Consultant/Engineer in charge. **(Basic rate of prepolished (one side) Granite @ Rs. 140/Sqft inclusive of all duties excluding taxes F.O.R. site)**.-For Kitchen. Rate is inclusive of providing & laying weather silicon of dowcorning make whereever required as ditected by Project Manager.

Providing and laying machine cut pre-polished stone slab in required size of approved quality in counters fixed over 19mm thick waterproof ply supported with MS angle frame with hilti anchor fasteners including providing and applying two or more coats of approved synthetic enamel paint over a coat of primer including surface preparation and including front facia, verticals, pointing with tinted white cement, polishing, curing etc. complete all as per drawings and as per detailed specifications, and joints neatly finished with white cement paste mixed with matching pigment to match the shade of stone making openings/ holes for wash basin, mixers, soap dispenser, ledge/ surrounding of bath tub, back splash, facia, ledges of shower area etc. as required including edge chamfering, moulding, mirror polishing of moulding and cut edges, and filling silicon at junctions with sanitary

Fixtures and corners with wall wherever required etc. complete as per drawing. Rate to include cost of necessary MS framework & basin cut size in ply. Plan area shall be measured, with all lead and lift and as directed by Project Consultant/Engineer in charge. -For Toilet. Rate is inclusive of providing & laying weather silicon of dowcorning make whereever required as directed by Project Manager.

With 18 to 20mm thick Prepolished Granite stone and joint to be filled with white cement slurry mixed with pigment to match the shade of stone **(Basic rate of republished (one side) Granite @ Rs. 140/Sqft inclusive of all duties excluding taxes F.O.R. site)**

Resource Deployment:
- Contractor shall deploy any or all the resource as dictated by the volume and nature of work.

Material:
- Fabricated angles 40X40x6mm with base plate 6mm thick
- Primer & synthetic enamel paint
- Approved Hilti Anchor Fastener
- 19 mm thick approved water proof plywood
- Approved adhesive.
- Nails
- Granite stone 18 to 20mm thick
- Synthetic Enamel paint
- P.P.E. As Required.

Tools & Tackless:

- Chisel.
- Hammer.
- hacksaw
- Grinder
- Brush
- Stone cutting machine.
- Grinding stone for stone edge polishing
- Safety PPE (Safety shoes, Helmet, Rubber Hand Gloves, Goggles, Nose Mask Etc.).

Preparatory Works:
- Surface should be cleaned to start work
- Ensure plywood for fixing of wooden ply
- Ensure angle should be fixed properly before laying stone.
- Ensure for primer coat is applied or not as per BOQ.
- Ensure for 2 coat synthetic enamel paint per mention in BOQ.

Application of Works:
- Marking of level as per drawing less 40mm from counter top to fix angle with 2 no's anchor fastener on base plate of angle.
- Cutting of Plywood as per drawing & fixing over angles with self-screw.
- Cutting of stone as per drawing.
- Pasting approved adhesive on plywood.
- Fixing of stone over plywood
- Chamfering &polishing on edges of stone.

Post check:

- After completion of work stone should be cleaned
- All corners of stone should be chamfered to avoid sharp edges.
- Level to be checked.

Material Testing Guidelines:
- Testing of material (Granit Stone, angle, plywood, primer, paint etc.) which used for fixing counter top shall be as per approved GPL Inspection Test Plan

Material Storage Guideline:
- The Granite slabs shall be stacked and tied on mils steel A frames
- The maximum slabs can be stacked each side of A frames should be 15 no's.
- Equate the stack weight on either side of the A frame during loading and unloading
- Care should be taken for any breakages during unloading and stacking the stone slabs.
- While stacking the stones, the polished surface should come in contact to each other
- Slabs of different quality, size and thickness shall be stacked separately

Safety Consideration:
- Each workman should use all PPEs like hand gloves, noose musk, safety goggles, safety belt, helmet and safety shoes during counter fixing works.
- All work will be instructed as per GPL work inspection.

Date

Project in Charge

Sub : Submission of Methodology of Gypsum boxing work.

Dear Sir,

Please find enclosed the copy of methodology for carrying out Gypsum boxing work at Godrej Nature + project.

For your record please.

<u>For Company Name</u>

Methodology of Gypsum boxing Work.

CLIENT-

Revision-00

Methodology of Gypsum boxing Work.

Objective.

- Objective of this procedure is to provide a guideline for sequence of operations pertaining to Gypsum boxing work, So as to ensure that the works are carried out in a systematic manner and to ensure that the works conform to the drawings and technical specification.

Scope.

- This procedure is applicable for Gypsum boxing work in structure.

REFERENCE.

- This Procedure is based on the requirements of the GFC Drawings and Technical Specification.

- **BOQ.**

1.6	False ceiling.	UNIT
	Providing and fixing suspended false ceiling which includes providing and fixing GI perimeter channels of size 0.55 mm. Thk. having one flange of 20 mm and another flange of 30 mm and web of 27 mm along the perimeter of the ceiling, screw fixed to brick wall/ partition with help of nylon sleeves and screws, at 610 mm centers. The suspended GI intermediate channels of size 45 mm,0.9mm thk with two flanges of 15 mm each from the soffit at 1220 mm centers with ceiling angles with ceiling angle of width 25mm x 10mm x 0.55mm thk fixed to soffit with GI cleat and steel expansion fasteners. Ceiling section of 0.55 mm thk having web of 51.5 mm and two flanges of 26 mm each with lips of 10.5mmare then fixed to the intermediate channels with help of connecting clips and in direction perpendicular to the intermediate channel with centers. All GI members sall be of Gypsteel or approved equivalent make. The 12.5 mm tapered edge gypboard (Confirming to IS-2095-1982) is then screw fixed to ceiling section with 25 mm dry wall screws at 230mm centers.	

Resource Deployment.

- The Contractor shall deploy any or all of the following resources as dictated by the volume and nature of work.

Material

- GI perimeter channels.

- Nylon sleeves.

- Screws.

- Ceiling section.

- 12.5 mm tapered edge gypboard.

- Intermediate channels

- Connecting clips

- Wall Fastener.

- Fiber tape.

- Fine Plaster of paris.

- P.P.E. As required.

Tools & Tackles

- Drill machine.

- Hammer

- Screw Driver.

- Ladder/Scaffolding.

- Safety PPE (Safety shoes, Helmet, Rubber Hand Gloves, Goggles, Nose Mask Etc.)
- Leveling Equipment.
- Nylon thread.

Preparatory Works.

- Proper Scaffolding need to arrange.
- Levelling need to be done for fixing perimeter and other channels.

Application of Works

- Proper level line need to fix with blue powder.

- Perimeter will be fixed on wall and ceiling with fastener & screw.

- Ceiling Section need to fix with proper gap to support gypsum board.

- Gypsum board need to be cut as per the frame & final sizes.

- Gypsum board will be fixed on the frame with proper screws.

- Fibre tape need to fix on the all the joints and corners.

- Very Fine plaster of Paris need to be paste on fibre tape & all the joints & screw halls.

Finish

- After completion of work check level maintain proper level of the Ceiling.

- Proper sanding need to be done on all the joints.

Tolerance

- Tolerance in the final finish could be 3 to 5mm (+) (-).

Safety Consideration and Requirement.

- Each workman should use all PPEs like hand gloves, nose musk, safety goggles, safety belt, helmet and safety shoes during Particular works.
- All work will be instructed as per GPL work inspection.

ACTIVITY: VACUUM DEWATERED FLOORING (VDF)

KNOWLEDGE OF ACTIVITY

A. DEWATERING AND FINISHING

- ➢ Concreting should be carried out with concrete of right grade and slump. (**See**: *Concreting)*

- ➢ Concrete intermediate/ alternate panels shuttered by ISMC runners or continuous bull marks.

- ➢ Ensure proper expansion joints as per working drawings and protected

- ➢ Dewatering mat to be used to remove excess water from the concrete surface to ensure concrete of the highest strength is achieved.

- ➢ The dewatering mat should be properly laid to prevent air leakage. Care should be taken in ensuring that excess water does not enter the dewatered area.

➢ After initial set of concrete (2-3 hours), the top surface of concrete is finished using a power trowel. Care should be taken that no impact load is transferred to the finished surface.

➢

➢ No load or stepping on concrete should be avoided, before initial set of finished concrete.

➢

➢ Heavy vibrations should be avoided in vicinity of the VDF area.

➢

B. TESTING OF CONCRETE

1.1   SLUMP TEST

➢ Conduct slump cone test, one during the start, one at midday and one towards the close of days work.

➢

➢ Any load under suspicion can be checked for slump on case-to-case basis.

➢

1.2   CUBE TEST

➢ The sample is made at point of discharge of mixture.

➢

➢ Take 3 samples after 7 days and 3 samples after 28 days strength.

> ➢
>
> ➢ Every 100 cum or 10 batches of concrete (whichever is smaller) should be represented by a sample of test cubes.
>
> ➢
>
> ➢ For concrete manufactured at site, there must be representative samples for the day's work and for different elements like footing, slab, column and wall.
>
> ➢
>
> ➢ Test results should be discussed with the project in-charge even if the results are satisfactory
>
> ➢
>
> ➢ In case of cube failure, the element to be identified and non-destructive test to be conducted.

TOOLS TO BE USED BY TRADESMEN

Required tools must be available at site to ensure correct work. Basic tools of the concreting gang are:

> ➢ Vacuum Dewatering machine
> ➢ Power Trowels
> ➢ Trowels
> ➢ Gum boots
> ➢ Dumpy levels
> ➢ Spirit levels 1-3m
> ➢ Measuring tape

Screed finishing Darby

➤ Steel trash tracks

➤ Hammer

➤ Leveling trowels

➤ Wheelbarrow

➤ Long handle Aluminum leveler with spirit level

➤ Masons brushes and buckets

➤ High pressure water cleaners

INSPECTION METHODOLOGY FOR QUALITY ASSURANCE

1. *Personnel*

➤ Ensure all personnel are wearing gumboots

2. *Before Concreting*

❖ Ensure shuttering is sealed with sealing tape so as there is no leakage of slurry

❖ All reinforcements are in place especially chairs

❖ All cover blocks damaged during reinforcement layout are replaced

3. *during Concreting*

❖ Ensure proper grade of concrete as recommended by the consultant is employed

❖ Ensure proper expansion joints as per working drawings and protected

❖ Check the top surface of fresh concrete for evenness of top surface

4. After Concreting

❖ Top surface is checked for straightness using Aluminium straight edge or leveling thread as the case may be.

❖ Curing carried as per guidelines

❖ Water level is maintained during ponding

❖ Ensure the cube test results are available at the end of the term indicated

TOOLS TO BE USED FOR QUALITY INSPECTION

✓ Dumpy levels to check finished slab level

✓ Aluminum straight edge

✓ Leveling thread

✓ Spirit levels 1-3m

✓ Measuring tape

✓ All the relevant related drawings

TECHNICAL SPECIFICATIONS FOR FINISHING WORK

1.0   GENERAL

1.1   **Scope**

This specification applies to the Civil, Structural, Finishing and External Development Works and building works to be executed by the Contractor. It is to be read in conjunction with and subject to the general conditions of contract and in conjunction with the drawings, the schedule of rates and such other documents as may from time to time be agreed upon as comprising part of this contract.   Where these specifications are not clear, relevant BIS codes and CPWD specifications shall be followed with prior permission of Engineer-in-Charge.

1.2 **Clearing**

The contractor shall clear the site of all rubbish and old buildings, remove all grass and low vegetation and remove all bush wood, trees, stumps of trees, and other vegetation only after consultation with the Engineer-in-Charge as to which bushes and trees shall be saved. All disused foundations, drains or other obstructions met with during excavation shall be dug out and cleared.

1.3 **Site Levels**

The contractor shall carry out the survey of the site and shall establish sufficient number of grids and level marks to the satisfaction of the Engineer-in-Charge, who shall decide on the basis of this information, the general level of the plot and the plinth.

1.4 **Bench-marks**

Prior to commencement of construction, the contractor shall in consultation with the Engineer-in-Charge, establish several site datum bench-marks, their number depending on the extent of the site. The bench-marks shall be sited and constructed so as to be undisturbed throughout the period of construction.

1.5 **Site investigation**

The Engineer-in-Charge might have got the soil investigation done and if so, the report will be handed over to the contractor for their scrutiny. The contractor shall however inspect the site and study the findings from the trial pits or bores in order to assess the problems involved in and methods to be adopted for excavation and earthwork. The contractor shall ascertain for himself all information concerning the sub-soil conditions, Ground water table periods and intensity of rainfall, flooding of the site and all data concerning excavation and earthwork. Any extra work required on this account, nothing will be paid extra.

1.6 Setting out the work

The contractor shall set out the works and during the progress of the building shall amend at his own cost any errors arising from inaccurate setting out.

During the execution of the work contractor must cross check his work with the drawings. The contractor shall be responsible for all the errors in this connection and shall have to rectify all defects and/or errors at his own cost, failing which the Engineer-in-Charge reserves the right to get the same rectified at the risk and cost of the contractor.

1.7 Cleaning up and handing over

Upon completion of the work all the areas should be cleaned. All floors, doors, windows, surface, etc. shall be cleaned down in a manner which will render the work acceptable to the Engineer-in-Charge. All rubbish due to any reason, shall be removed daily from the site and an area of up to ten metres on the outer boundaries of the premises will be cleaned by the contractor as a part of the contract. Upon completion of the project, the contractor shall turn over to the Engineer-in-Charge the following:

a)   Written guarantee and certificates.

b)   Maintenance manuals, if any, and

Keys.

1.8   **Samples**

The contractor shall submit to the Engineer-in-Charge samples of all materials for approval and no work shall commence before such samples are duly approved. Samples of precast concrete panels, masonry units, building insulation, finished hardware, metal window and door frames, terrazzo flooring, kota stone, marble etc. and every other work requiring samples in the opinion of the Engineer-in-Charge shall be supplied to the Engineer-in-Charge, and these samples will be retained as standards of materials and workmanship. The cost of the samples shall be borne by the contractor.

Throughout this specification, types of material may be specified by manufacturers' name in order to establish standard of quality, price and performance and not for the purpose of limiting competition. Unless specifically stated otherwise, the tenderers may assume the price of 'approved equivalent' except that the burden is upon the contractor to prove such equality, in writing.

A detailed programme shall be submitted by the Contractor for the material approvals, within four weeks of the Engineer-in-Charge's order to commence. The detailed programme shall include but not limited to:

Date/s of submitting the various material samples.

Date/s by which the Engineer-in-Charge's approval is required.

Date/s of placing orders on the Manufacturers/Suppliers.

Date/s of arrival of the approved material/s on to the site.

Date/s of the completion of the `Mock-ups', wherever required, and the Date/s by which the Engineer-in-Charge's inspection of such `Mock-ups' should be completed and the Date/s by which the Engineer-in-Charge should fully approve the said Mock-ups.

1.9 Tests

All materials and methods of tests shall conform to the latest rules, regulation and/or specifications of the following authorities where specified herein as applicable. Bureau of Indian Standards (BIS), British Standards Code of Practice (BS) in case no equivalent BIS is available. The Engineer-in-Charge will have the option to have any of the materials tested and if the test results show that the materials do not conform to the specifications, such materials shall be rejected. A reasonable number of representative tests will be deemed to be included in the rates tendered.

1.10 Mode of Measurements

All measurements will be taken in accordance with IS 1200 latest issue unless otherwise specified.

1.11 Cost of activities under Para 1.1 to 1.9, the contractor should include in their quoted rate as well as contract scope of work may be considered.

2.0 WOOD WORKS

The Contractor shall be responsible for providing all plant, tools, materials, labour and all things necessary for the proper execution, completion and maintenance of the works.

a) **Timber :-**

The moisture content of the timber during manufacture, delivery to site, storage, site working, assembly, installation shall be 10 to 12 percent.

Timber shall be Burma Teak Wood/ Ivory Coast Teak Wood / Champ Wood, / Red mirinti, soft or hardwood and shall be suitable for the purpose for which it is intended. It shall be seasoned or Kiln dried, absolutely free from worm holes, large loose or dead Knots or other defects which would effect strength or usability and shall be flat, straight non-splitting and dressed on all sides. It shall be matched for colours and graining.

Burma Teak Wood/ Ivory Coast Teak Wood / Champ Wood / Red mirinti wherever specified in the drawings / schedule of quantities , it is Ist quality Light Grained of reasonably straight grains, light vein free of Knots and sap.

Fixing :-

The carpentry timber shall be fixed with nails, spikes, bolts screws, hangers, stirrups, anchors, ties or any other accessories which are suitable to develop the full strength of the member to which they are attached, as directed.

Carpentry timber where fixed to solid masonry or concrete shall be secured with expansion bolts or other positive methods of mechanical fastening. MS hold fast grouted in CC block shall be used to hold the door frames.

Timber - Treatment

All timber shall be protected with an organic solvent water repellent wood preservative to give a highly efficient protection against termite, spider, worm, all insects and fungus and rot attach and shall, where exposed, enhance the appearance of the timber. Colour of the product shall be such as to bring out the natural colour of the respective timbers. Fire retardant paint to timber shall be applied as per the recommendations of manufacturer and shall comply with the requirement of ISI / B.S. code and local fire requirements.

b)   Plastic Laminate :-

Plastic decorative laminate sheeting shall be of the brand, catalogues number and indicated or approved. Plastic laminate shall be fire retardant to class I of BS : 476 or ISI code where specified.

c)   Veneers :-

Veneer shall be of the timber species of Indian origin shown on drawings. Veneers are to be kept in sequence as they are being cut from wood and supplied as such to the site for accurate matching or figuring . The veneer shall be finished as specified and shall be equal or superior quality to that laid down in IS: 1659 - 1960 or as approved.

wherever Veneer is stated, it is mainly Teak veneer of approved quality in Indian origin unless otherwise mentioned.

d)   Plywood :-

All ply wood shall be of best / high quality close grained suitable for veneering, painting or bonding plastic laminate. It shall be resin bonded and weather proof. Exposed edges shall be finished with an edge strip of solid teak wood. tongued and grooved and glued , or as detailed. The plywood of approved brand and manufacture only shall be used in the work. The thickness shall be in accordance with the drawings /schedule of quantities.

e)   MDF (Medium Density Fibre Board):

For Interior Works MDF of approved make /manufacturer shall be of only exterior grade as per IS : 12406 -1988. It is to be contained that MDF shall be invariably used in place of Ply / Boards .so specified in the specifications of either same thickness or of higher thickness .wherever feasible The minimum thickness of MDF to be used shall be 8mm. Wood screws are not to be used for MDF and only fully threaded parallel shank screws shall be used after drilling pilot holes. Veneering /lamination to the MDF surface shall be done by exterior grade adhesive only. Poly urethene primers shall be used for sealing the edges and painting the rear side. For specifications of various applications the manufacturer users manual shall be followed.

Adhesives :-

The adhesives used for all wood work and MDF shall be FEVICOL or approved equivalent of appropriate Grade. Manufacturer's recommendations shall be followed for adhesive other than above required for any specified / specialized work.

Joinery :-   Joinery shall be carried out strictly in accordance with the drawings, Where joints are not specifically indicated recognized forms of joints shall be used. Joinery shall conform to IS Standards.

Panels shall be rendered flame retardant and to conform to local fire regulations.

The Contractor shall submit samples of all materials including samples of veneer for approval.

All materials pre-fabricated, delivered and assembled shall be in accordance with the approved sample.

The Contractor shall be responsible for protecting all items of wood-work done by him. The contractor shall replace at his own expense any damaged work caused through lack of adequate protection or care in installation or handling .

f)   Flush Doors

All flush door shall be solid core as specified. It shall conform to the relevant specifications to IS: 2202 and shall be obtained from ISI approved manufacturers. The finished thickness of the shutter shall be as mentioned in the items. Face veneers shall be of the pattern and colour approved by the Architects and an approved sample shall be deposited with the Architects for reference. The solid core shall be wood laminates prepared from battens of well seasoned and treated good quality wood having straight grains.

g)   Gypsum - Board :-

Gypsum -Board (Glass Fibre Reinforced Board) or Equivalent conforming to IS-2095 - 1982 and 2542-1981 shall be used . Technical detailing for fixing Gypsum-Board along with jointing compound, paper tape, primer, screws, edge bead, angle bead etc. shall be as per Manufacturers specification. Proper care is to be taken while handling, storing and cutting the Gypsum-board as per manufacturer's manual and the work shall be done in technical co-ordination /assistance with the trained staff of Manufacturer, such services being offered free by them.

h)   Mirrors and Glasses :-

Mirrors shall be fabricated from best clear plate or float glass of approved quality in imported variety and shall match the International Standards. All fixed panel mirrors shall be +/- 0.30mm tolerance . The edges of mirrors shall be polished and bevelled and mitred as per IS specifications wherever, it's indicated in the drawing.

All vision glasses shall be float glass of specified thickness. The edges shall be bevelled as indicated in drawings and shall be done at approved source.

The Etching wherever specified in drawings, shall be done at approved sources as per full-scale drawing approved by the Project Manger.   The etched panel shall be chemically washed /treated as per specialist specifications to have a permanent dust free surface.

The Contractor shall be responsible for protecting all mirrors and glasses fixed by him and shall replace at his own expense any broken or damaged mirror / glass caused through lack of adequate protection or care in installation or handling.

l)   Rough Ground :-

Providing and fixing rough ground of approved size as per the elevation drawings, fabricated out of heavy duty aluminium extruded profiles anodized / Powder Coated as profiles in approved shade including fasteners, screws etc as required, providing Masking Tapes on the profiles for safety against external scratches at site (Masking Tapes to be removed only at the time of installation of aluminium doors and windows) complete as per the instructions of Engineer in Charge.

Materials Specification :- All materials and finishes are to be new and free from defects which may impair the appearance, strength, function and durability of the exterior window system and related construction of the external coverings.

Aluminium: The aluminium-extruded sections shall conform IS designation HE/HV/9WP alloy with chemical composition and mechanical properties as per IS 733:1975 wall thicknesses to meet required loadings, with minimum for trim being 2.6mm. Test certificate for alloy and its extrusion from the manufacturer is required to be submitted by the contractor for its conformity.

Coating/ Anodising: All aluminium sections shall be anodised or powder coated. Anodising shall conform to IS: 1868 -1982 and shall be of AC 25 grade with minimum thickness of 25 +/- 3 microns when measured as per IS 6012-1992 and the density shall be at least 32mg/square inch. All sections are to be matt anodised in colour as per sample available with the architects. The anodic coating shall be properly scalled by steam or boiling in de-ionised water as per IS 1868-1982. In case of powder coating, factory applied electrostatic powder coated sections 60+/- 5 micron will be considered for approval. Colour consistency shall be accurate. Abrasion Resistance shall confirm to IS – 5523 – 1983

No visual variation in shade shall be permitted. The fabricator shall clearly indicate the shade variation tolerance as measured by standard equipme Density of Aluminium Section :- 2700Kg/cum The following section shall be used :- Over All Size 50mmx25mmx1.5mm - 0.6075 Kg/Rmt Over All Size 50mmx15mmx1.5mm - 0.5265 Kg/Rmt OR approved section.

3.0 FLOORING/ CLADDING WORKS

3.1 General

All flooring shall be laid to the best practice known to the trade. The flooring shall be laid to the level except where slopes are called for on the drawings in which case the slopes shall be uniform and so arranged to drain in to the indicated outlets.

Particular care shall be exercised to ensure that all flooring, skirting and dado are perfectly matched for colour and finish. Sufficient extra tiles (not less than 5%) shall be cast/ordered to ensure an adequate supply of matched floor tiles. The contractor shall furnish for approval by the Engineer-in-Charge, samples of each type of floor finish.

3.2 Cement Concrete flooring (IPS Flooring)

Indian patent stone flooring shall be 40mm or of specified thickness and laid in two layers, bottom layer 28mm thick or as specified in 1 part of portland cement, 2 parts of coarse sand and 4 parts of crushed stone aggregate 12.5mm down well graded machine mixed with not more than 5.5 gallons of water for each bag of cement and top layer 6mm thick in one part of portland cement, 2.5 parts of selected crushed stone chips with just enough sand maximum part to make workable mix, machine mixed with not more than 5 gallons of water. Top layer to be laid before the bottom layer has hardened. Flooring shall be laid in squares or bays as directed and each layers shall be well compacted by ramming with heavy teak wood floats. The top shall be brought to a smooth and even surface free from blemishes and finished smooth with neat cement by steel trowelling. The flooring shall be kept wet for seven days for curing. The flooring shall be laid by forming panels as per the advised of the Engineer-in-Charge / Architect including roughening of the base layer and providing cement slurry for jointing.

Where ironite/hardonite topping is specified in the "Bill of Quantities" the bottom layer shall be 50mm thick or in the item of B.O.Q. and the top layer shall be 12mm thick mixed with ironite/hardonite as per manufacturers specification and finished fair.

3.3 Granolithic Flooring

The general specifications for granolithic floors, where called for, shall be as per the cement flooring except that the top 12mm finish shall be of granolithic consisting of 1 part of cement and 1.1/2 part of well graded crushed aggregate. The aggregate shall be of approved quality.

3.4 Tiles

All tiles shall be minimum 8 mm thick of approved manufacturer as stated in the schedule of quantities. Only first quality tiles of approved colour shall be used. No cracked or warped tiles shall be used in the work. All tiles shall be required to be set in cement mortar. Prior to setting the tiles the contractor shall at his own cost, clear the whole surface and thoroughly saturate it with water. A layer of 12 to 20 mm avg. thick cement mortar shall then be applied to the surface and the tiles laid firmly over a layer of clear cement slurry. The tiles shall be set in perfect line, level and true to plumb line. The joints of tiles shall have white, coloured cement filling/tile grout. After the setting operation is completed, the contractor shall carefully remove all cement and dribbling and cure the tiled surface for atleast seven days with water.

3.5 Tile Dado

Tile dado where called for in the drawings, shall be minimum 6 mm thick tiles of approved manufacture. The tiles shall be free from cracks, twists, uneven edges, cracking and such other defects. The rear face of tiles shall be grooved and/or recessed to provide an adequately key for the plaster. A layer of 12mm thick rough base plaster shall be done with cement mortar 1:3   (1 cement: 3 coarse sand). The tiles shall be finally set true to plumb with rich cement slurry over a 6mm average thick cement plaster in 1:3 (1cement : 3 fine sand), the joints of tiles shall have white, coloured cement filling/tile grout. After laying the tiles shall be thoroughly washed and cleaned to the satisfaction of the Engineer-in-Charge.

3.6 Kota Stone Flooring

The best quality stone (Green /Brown) from approved quarry, shall be laid either with rough stone or machine cut and machine polished as specified in respective items and shall be of specified thickness and of approved quality and size, free from cracks and flakes and shall be uniform in colour, with straight edges. The sides of machine cut and machine polished stones shall have perfect right and finished. The stones shall be laid on minimum thickness of 25mm thick cement mortar 1:4 (1 cement : 4 coarse sand) mix to match the total thickness of flooring of 50mm thick & joint to be fitted with cement slurry mixed with pigment to match the shade of stone. The finished stone surface including edge prepared surface thus laid shall then be polished to the required degree as approved by the Engineer-in-Charge. Flooring shall be finally mirror polished and protected till the handing over of the building.

3.7 Marble / Granite Stone

Marble/Granite shall be the best Indian Marble/Granite to be approved by the Engineer-in-Charge and a sample piece should be kept in the office of the Engineer-in-Charge. The quality shall be uniform and it shall be hard and free from any discolorations, cracks, flaws, veins of foreign materials or any other defects. When marble/Granite of different colour and kinds associated, care shall be taken to see that they are of equal hardness so as to wear evenly. The marble/Granite slabs shall be machine cut true to the shape and size and machine mirror polished. Care shall be taken to cut the slabs so as to provide a pattern as indicated. Marble/Granite stone slabs for wall lining and dadoes shall be machine mirror polished edges. The wall shall be lined with the marble/Granite in courses as indicated and grain of the marble/Granite shall be arranged in pattern as per detailed drawings. The marble/Granite shall be bedded in adequate thickness of cement mortar, backing covering the full area of the marble. The wall surface shall cleaned from all dirt, mortar droppings etc. before applying the base plaster. The marble/Granite shall be fixed to the wall by S.S cramps and pins of required sizes embedded firmly in to wall by cutting hole and grouting alternately stainless steel cramps and pins as per design including fixing small stone pieces with adhesive. Fixing of cramp shall be with fastener (as approved by Engineer-in-Charge) in case of RCC and in Brick Work pocket (100mmX75mmX75mm) shall be made and filled by non-shrinkage compound (as approved by Engineer-in-Charge). The size and design of the cramp and fastener shall be to suite site requirement and shall be approved by the Engineer-in-Charge. The load of one marble/Granite slab shall not be borne by the slab below. Joints between slabs shall be hair fine and filled with coloured cement to match the marble/Granite. The marble/Granite lining and dadoes shall be finally polished by Carborundum stone, buffing with polishing felt and cleaned with diluted oxalic acid wash.

3.8   Expansion and compression joints

These shall be clearly indicated on the shop drawings and formed of non-staining two parts polysulphide with polyethylene foam backing to full depth of screed in pavings.

In no instance shall expansion joints be less than 10 mm. Supporting corbels cover shall be recessed into the back of the above slab and not placed in the expansion joint. Expansion joint shall be kept completely free of all fixing materials and are to be inspected by the consultant prior to filling.

1.2.1.1 3.9   VACCUM DEWATERED CONCRETING/TREMIX FLOORING

PREPARATION -

The surface to receive flooring shall be clean, free from dirt and free from foreign material.

Any undulations or mortar remaining on the floor shall be trimmed.

Base course shall be trimmed.

The base shall be cleaned and watered before laying the floor.

Work includes at all depths and heights. The finished surface shall be kept wet for a maximum period of one week.

CONCRETING

General

Concreting shall have a concrete base of M20 of specified thick.

Flooring shall have hard top on the concrete base.

Flooring shall be laid in strips, the size of which is mentioned on the drawings.

Materials

Cement   -   Portland

Sand   -   River sand

Aggregate   -   Max. size 10 to 20mm

Water   -   Potable

Floor hardener - @3kG/Sqm

Execution -

Mix cement, sand and aggregates as per grade M20 thoroughly with water to get an appropriate consistency. Prepared concrete shall be laid immediately after mixing. The base shall be free from water and other foreign materials, dust and dirt. A coat of cement slurry of the consistency of thick cream shall be brushed on the surface of the base course. The concrete shall then be spread over this base evenly and leveled carefully. Low areas shall be filled with concrete and humps removed. Devacumisation shall be done for removing the voids.

1. The whole concrete surface shall be leveled, compacted by ramming and trowelling.
2. Prepared surface shall be allowed to set.

3. **Hardner screed**
4. Hard top to be prepared as per the specifications with Nitohardner and one part of dry cement.
5. The hard top shall be provided over concrete base immediately after it is set, compacted and leveled with a steel trowel.

6. The surface shall be trowelled to bring the hardener coat to a leveled surface.
7. Excessive trowelling shall be avoided.
8. After the initial set, further compaction shall be done by steel trowelling.
9. Final brushing shall be made before the floor top becomes too hard.
10. Provision for expansion joint to be made as shown in the drawing.

CURING

11. Curing shall commence as soon as the surface is hard enough to receive the water.
12. The surface shall be covered with sacks or sand and shall be kept continuously wet for a period of at least one week.

4.0 FINISHING WORKS

4.1 General

4.1.1 All plaster work shall be of the best workmanship and in strict accordance with the dimensions of the drawings. All plastering shall be finished to true levels including plumbs, without imperfections, and square with adjoining work. It shall form proper foundations for finishing materials such as paint etc. Masonry and concrete surface to which plaster is to be applied shall be clean, free from efflorescence, sufficiently rough and keyed to ensure proper bond.

4.1.2 Wherever directed all joints between RCC frames and masonry walls, shall be expressed by a groove in the plaster. This groove will exactly coincide with the joint beneath. At the corners of all windows and doors or other openings and wherever instructed, 24 gauge expanded galvanized metal mesh strips 300 mm wide shall be placed diagonally to prevent plaster cracks.

4.1.3   Where grooves are not called for, the joint between concrete and masonry in filling, chasing for conduits, pipes, boxes etc. shall be covered by 24 gauge expanded galvanized metal strips, 300 mm wide installed before plastering. The contractor shall supply all necessary labour, material, tools and scaffolding necessary for the completion of the work detailed. He shall be responsible to take proper precautions to all works from damage. Any work rejected through non-compliance with the specifications or damaged work shall be removed and replaced at the expense of the contractor.

4.1.4   All chasing, installation of conduits, boxes, etc. shall be completed before any plastering is commenced on a surface. Chasing or cutting of plaster will not be permitted. Broken corners shall be cut back less than 150 mm on both sides and patched with plaster of Paris as directed. All corners shall be rounded to a radius. Contractor shall get samples of each type of plaster work approved by the Engineer-in-Charge.

4.1.5   The materials used for plastering shall be proportioned by volume by means of gauge boxes. Alternatively it may be required to proportion the materials by weight.

4.2   Plaster Work

4.2.1   The joints in the brick work, concrete blocks, shall be raked (as specified in respective subhead) while the masonry is green. Concrete surfaces to receive plaster shall be suitably roughened. All walls shall be washed with water and kept damp for 10 hours before plastering.

4.2.2   The plaster unless specified otherwise shall be average of 15 mm thick on walls and minimum 6 mm thick for the ceiling. The finished texture shall be as approved by the Engineer-in-Charge. The mix for plaster unless otherwise specified, shall be one part cement and four parts sand, to walls and one part cement, 3 parts sand to ceiling.

4.2.3   The interior plaster shall be applied in one coat only. The surface shall be trowelled smooth to an approved surface. All plaster work shall be kept continuously wet for seven days.

4.2.4   The external plaster shall be minimum 20 mm. Preparations of walls to receive plaster work shall be the same as in internal plaster. Both layers of all external plaster shall be waterproofed with approved water proofing powder added to cement in proportion of 1.5 Kg. to 50 Kg. of cement as per the manufacturers' instruction, for both the coats under layer 12 mm thick cement plaster 1:5 (1 cement : 5 coarse sand) finished with a top layer 8mm thick cement plaster 1:6 (1 cement :6 fine sand) including making of grooves, bands, drip coarse as shown in drawing.

4.2.5   For sand faced cement plaster, the finishing coat shall be in cement mortar 1:3, sand used shall be of selected colour, properly graded and washed so as to give a grained texture. Finishing plaster coat shall be 8 mm thick, uniformly applied and surface finished with special rubbing by sponge pads and other tools and recommended by the Engineer-in-Charge.

4.3   White Washing

4.3.1   White washing with Lime

The wash shall be prepared from fresh stone lime (Narnaul/Satna or Dehradun quality). The lime shall be thoroughly slaked on the spot, mixed and stirred with sufficiencies to water to make a thin cream. This shall be allowed to stand for a period of 24 hours and then shall be screened through a clean coarse cloth. 40 gm of gum desolved in hot water, shall be added to each 10 entire delimiters of cream. The approximate quantity of water to be added in making ht cream will be 5 liters of water to 1 Kg. of lime.

Indigo (Neel) up to 3 gm. per Kg. of lime desolved in water, shall then be added and wash stirred well. Water then shall be added at the rate of about 5 liters per Kg. of lime to produce a milky solution.

4.3.2 Preparation of surface

Before white washing is started, the surface shall be thoroughly brushed free from mortar droppings and foreign-matter. Any unevenness shall be made good by applying putty made of plaster of Paris mixed with water on the entire surface including filling up the undulations and then sand papering the same after it dry.

4.3.3 Application

The white wash shall be applied with moon brushes to the specified number of coats. The operation for each coat shall consist of a stroke of the brush given from top downwards, another from bottom upwards over the first stroke, and similarly one stroke horizontally from the right and another from the left before it dries up.

4.4 White washing with whiting

Preparation of mix : Whiting (ground white chalk) shall be dissolved in sufficient quantity of warm and thoroughly stirred to form a thin slurry which shall then be screened through a clean coarse cloth. Two Kg. of gum and 0.4 Kg. of copper sulphate dissolved separately in hot water shall be added for every cum of the slurry which shall then be diluted with water to the consistency of milk also as to make a wash ready for use.

Other specifications described in above shall be applied in this case also.

4.5   Colour Washing

The mineral colours not affected by lime, shall be added to white wash. Indigo shall however, not be added. No colour wash shall be done until a sample of the colour wash of the required tint or shade has been got approved from the Engineer-in-Charge. The colour shall be of even tint or shade over the whole area.

A priming coat of white wash with lime or with whiting shall be applied. Three or more coats, shall then be applied on the entire surface till it represents a smooth and uniform finish.

Other specifications described in above shall apply in this case also.

4.6   Distempering

Dry distemper of required colour and (IS; 427 - 1965) of approved brand and manufacture shall be used. The shade shall be got approved from the Engineer-in-Charge before application of the distemper. The dry distemper colour as required shall be stirred slowly in clean water using 6 decilitres (0.6 litre) of water per Kg. of distemper or as specified by the makers. Warm water shall preferably be used. It shall be allowed to stand for atleast 30 minutes (or if practicable over night) before use. The mixture shall be well stirred before and during use to maintain an even consistency. Distemper shall not be mixed in larger quantity than is actually required for one days' work.

Preparation of surface

Before new work is distempered, the surface shall be thoroughly brushed free from mortar droppings and other foreign matter and sand papered smooth. Pitting in plaster shall be made good with plaster of Paris mixed with the colour to be used. The surface shall then be rubbed down again with a fine grade sand paper and made smooth. A coat of distemper shall be applied over the patches. The patched surface shall be allowed to dry thoroughly before the regular coat of distemper is applied.

A priming coat of whiting shall be applied over the prepared surface. No white washing coat shall be used as a priming coat for distemper.

Application

The treatment shall consist of a priming coat of whiting followed by the application of three or more coats of distemper till the surface shows an even colour.

Other specifications described as above shall apply in this case also.

4.7   Oil bound distempering

Material: Oil emulsion (oil bound) distemper (IS:428-1929) of approved brand and manufacture shall be used. The primer used shall be cement primer or distemper primer. This shall be of same manufacture as distemper. The distemper shall be diluted with water or any other prescribed thinner in a manner recommended by the manufacturer. Only sufficient quantity of distemper required for days work shall be prepared. The distemper and primer shall be brought by the contractor in sealed tins in sufficient quantities, at a time to suffice for a fortnights work. The empty tins shall not be removed from the site of work, till this item of work has been completed and passed by the Engineer-in-Charge.

Preparation of surface

Before new work is distempered, the surface shall be thoroughly brushed free from mortar droppings and other foreign matter and sand papered smooth. Pitting in plaster shall be made good with plaster of Paris mixed with the colour to be used. The surface shall then be rubbed down again with a fine grade sand paper and made smooth. A coat of distemper shall be applied over the patches. The patched surface shall be allowed to dry thoroughly before the regular coat of distemper is applied.

A priming coat of approved primer shall be applied over the prepared surface. No white washing coat shall be used as a priming coat for oil bound distemper.

Application

The priming coat shall be with cement primer, as required in the description of the item and as recommended by the manufacturer.

Note:

If the wall surface plaster has not dried completely cement primer shall be applied before distempering the walls. But if distempering is done after the wall surface is dried completely, distemper primer shall be applied.

Oil bound distemper is not recommended to be applied within six months of the completion of wall plaster.

After the primer coat has dried for atleast 48 hours, the surface shall be lightly sand papered to make it smooth for receiving the distemper, taking care not to rule out the priming coat. All loose particles shall be dusted off after rubbing. One coat of distemper properly diluted with thinner (Water or other liquid as stipulated by the manufacture) shall be applied with brushes in horizontal strokes followed immediately by vertical ones which together constitute one coat. The subsequent coats shall be applied in the same way.

For distemper 15 cm double bristled brushes shall be used. After each days work, brushes shall be thoroughly washed in hot water with soap solution and hung down to dry. The final coat shall be done by using approved roller.

The specifications in respect of scaffolding protective measures and rute shall be as described under.

4.8 Cement Primer Coat

Cement primer shall be used as lease on wall finish of cement lime or lime cement plaster or asbestos cement surface before oil distemper paints are applied on them. Only approved cement primer shall be used. Primer coat shall be preferably applied by brushing and not by spraying.

Preparation of surface

The surface shall be thoroughly cleaned of dust, old white or colour wash by washing and scrubbing. The surface then be allow to dry for atleast 48 hours. It shall then be sand papered to give a smooth and even surface. Any unevenness shall be made good by applying putty, made of plaster of Paris mixed with water on the entire surface including filling up the undulations and then sand papering the same after it is dry.

Application

Cement primer shall be applied with a brush. Horizontal strokes shall be given first and vertical strokes shall be applied immediately afterwards. The entire operation will constitute one coat. The surface shall be finished as uniformly as possible leaving no brush marks. It shall be allowed to dry for atleast 48 hours, before oil emulsion paint is applied.

4.9 Cement Paint

Cement paint shall be (conforming to IS:5410 - 1969) of approved brand and manufacture.

Preparation of surface

The surface shall be thoroughly cleaned of all mortar dropping, dirt, dust, alga, grease and other foreign matter by brushing and washing. The surface shall be thoroughly wetted with clean water before the cement paint is applied.

Preparation of mix

Cement paint shall be mixed in such quantities as can be used up within an hour of its mixing as otherwise the mixture will set and thicken, affecting flow and finish.

Cement paint shall be mixed with water in two stages. The first stage shall comprise of 2 parts of cement paint and one part of water stirred thoroughly and allowed to stand for 5 minutes. Care shall be taken to add the cement paint gradually to the water and not vice versa. The second stage shall comprise of adding further one part of water to the mix and stirring thoroughly to obtain a liquid of workable and uniform consistency. In all cases the manufacturers instructions shall be followed meticulously. The lid of cement paint drums shall be kept tightly closed when not in use, as by exposure to atmosphere the cement paint rapidly becomes air set due to its hydrophobic qualities.

Application

The solution shall be applied on the clean and wetted surface with brushes or spraying machine. The solution shall be kept well stirred during the period of application. It shall be applied on the surface which is on the shady side of the building so that the direct heat of the sun on the surface is avoided. The method of application shall be as per manufacturer's specifications. The completed surface shall be watered after day's work.

Water cement paint shall not be applied on surface already treated with white wash, colour wash distemper dry or oil bound, varnishes, paints etc. It shall not be applied on gypsum, wood and metal surfaces.

4.10   Painting

i)   **Painting priming coat of wood surface**

Primer for wood work shall be as specified in the description of the item. Surface to be primed shall be dry and thoroughly cleaned. All unevenness shall be rubbed down smooth with sand paper and shall be well dusted, knots, if any, shall be covered with preparation of red lead made by grinding red lead in water and mixing with strong glue sized and used hot. Appropriate wood filler material with same shade as paint shall be used where so specified.

The surface treated for knotting shall be dry before primer is applied. After the primer is applied the holes and indentation on the surface shall be stopped with glaziers putty or wood putty, stopping shall not be done before the priming coat.

ii)   Painting priming coat on Iron & Steel surfaces

All rust and scales shall be removed by scrapping or by brushing with steel wire brushes. Hard skin of oxide formed on the surface of wrought iron during raking which becomes loose by rushing, shall be removed. All dust and dirt shall be thoroughly wiped away from the surface.

iii)   Textured paint

The textured finish to external surfaces of walls as per manufacturer's specification and approved by the Engineer-in-Charge including scaffolding etc. complete.

iv)   Painting priming coat on plastered surface

The surface shall ordinarily not be painted shall be applied to get correct finish until it has dried completely. Before primer is applied, holes and undulations shall be filled up with plaster of Paris and rubbed smooth.

The primer shall be applied with brushes, worked well into the surface and spread even and smooth. Painting shall be done by crossing and laying off. The crossing and laying off consists of covering the area over with paint, brushing the surface hard for the first time over and then brushing alternately in opposite direction, two or three times and then finally brushing lightly in a direction at right angles to the same. In this process, no brush marks shall be left the laying off is finished. The full process of crossing and laying off will constitute one coat.

The surface to be painted shall have received the approval of the Engineer-in-Charge after inspection, before painting is commended.

Application

The number of coats including the under coat shall be stipulated in the item.

a) Under Coat

One coat of specified paint of shade suited to the shade of the top coat shall be applied and allowed to dry overnight. It shall be rubbed next day with the finest grade of wet abrasive paper to ensure a smooth and even surface, free from brush marks and all loose particles dusted off.

b) Top Coat

Top coats of specified paint of desired shade shall be applied. Each coat shall be allowed to dry for not less than 24 hours and lightly rubbed down smooth with finest wet abrasion paper to get an even glossy surface. If, however, the surface is not satisfactory additional coats as required.

4.11 Acrylic Emulsion paint

The acrylic emulsion paint is not suitable for application on external wood, and iron surface and surfaces which are liable to heavy condensation and are to be used on internal surfaces except wood and steel which are liable for condensation. No priming coat is required for the latter.

Acrylic emulsion paint as per IS : 5411 – 1969 of approved brand and manufacture and of the required shade shall be used.

4.11.1   Painting on New Surface

Preparation of surface

Before new work is distempered, the surface shall be thoroughly brushed free from mortar droppings and other foreign matter and sand papered smooth. Pitting in plaster shall be made good with plaster of Paris mixed with the colour to be used. The surface shall then be rubbed down again with a fine grade sand paper and made smooth. A coat of paint shall be applied over the patches. The patched surface shall be allowed to dry thoroughly before the regular of acrylic emulsion paint is applied.

A priming coat shall be applied over the prepared surface. No white washing coat shall be used as a priming coat.

Application

The number of coats shall be as stipulated in the item. The paint will be applied in the usual manner with brush and roller. The paint dries by evaporation of the water content and as soon as the water has evaporated the film gets hard and the next coat can be applied. The time of drying varies from one hour on absorbent surfaces to 2 to 3 hours on non-absorbent surfaces.

The thinning of emulsion is to be done with water and not with turpentine. Thinning with water will be particularly required for the undercoat which is applied on the absorbent surface. The quantity of water to be added shall be as per manufacturer's instructions.

The surface on finishing shall present a flat velvety smooth finish. If necessary more coats will be applied till the surface presents a uniform appearance.

Precautions

    a)   Old brushes and rollers if they are to be used with emulsion paints, should be completely dried of turpentine or oil paints by washing in warm soap water.

            Brushes and rollers should be quickly washed in water immediately after use and kept immersed in water during break periods to prevent the paint from hardening on the brush and roller.

    b)   In the preparation of walls for plastic emulsion painting, no oil base putties shall be used in filling cracks, holes etc.

    c)   Splashes on floors etc. shall be cleaned out without delay as they will be difficult to remove after hardening.

    d)   Washing of surfaces treated with emulsion paints shall not be done within 3 to 4 weeks of application.

4.12 Synthetic Enamel Paint

Synthetic Enamel Paint (conforming to IS : 1932 – 1964) of approved brand and manufacture and of the required colour shall be used for the top coat and an undercoat of shade to match the top coat as recommended by the manufacturer shall be used.

4.12.1 Painting on New Surface

Preparation of surface

a) Wood work :- The surface shall be cleaned and all unevenness removed. Knotes if available, shall be covered with a preparation of red red lead. Holes and indentations on the surface shall be filled in with glazire's putty or wood putty and rubbed smooth before painting is done. The surface should be thoroughly dry before painting.

b) Iron and steel work :- The priming coat shall have dried up completely before painting is started. Rust and scaling shall be carefully removed by scraping or by brushing with steel wire brushes. All dust and dirt shall be carefully and thoroughly wiped away.

c) Plastered surface:- The priming coal shall have dried up completely before painting is started. All dust and dirt that has settled on the priming coat shall be thoroughly wiped away before painting is started.

Application

The number of coats including the undercoat shall be as stipulated in the item.

Under Coat :- Two coat by spray method (by using compressor) of the specified paint of shade suited to the shade of the top coat shall be applied and allowed to dry overnight. It shall be rubbed next day with the finest grade of wet abrasive paper to ensure a smooth and even surface, free form brush marks and all loose particles dusted off.

Top Coat :- Top coats of specified paint of the desired shade shall be applied by spray method (by using compressor) after the undercoat is thoroughly dry. Additional finishing coats shall be applied if found necessary to ensure properly uniform glossy surface.

4.13 Polishing New Surface

Preparation of surface

a) The surface shall be cleaned. All unevenness shall be rubbed down smooth with sand paper and well dusted. Knots if visible shall be covered with a preparation of red lead and glue size laid on while hot. Holes and indentations on the surface shall be stopped with glazier's putty. The surface shall then be given a coat of wood filler made by mixing whiting (ground chalk) in methylated spirit at the rate of 1.5kg of whiting per litre of spirit. The surface shall again be rubbed down perfectly smooth with glass paper and wiped clean.

Application

The number of coats of polish to be applied to achieve the desired shade / finish.

A pad of woollen cloth covered by a fine cloth shall be used to apply the polish. The pad shall be moistened with the polish and rubbed hard on the wood, in a series of overlapping circles applying the mixture sparingly but uniformly over the entire area to given an even level surface. A trace of linseed oil on the face of the pad facilitates this operation. The surface shall be allowed to dry and the remaining coats applied in the same way. To finish off, the pad shall be covered with a fresh piece of clean fine cotton cloth slightly damped with methylated spirit and rubbed lightly and quickly with circular motions.

The finished surface shall have a uniform texture and high gloss.

4.14   Melamine Polish

Timber works shall be finished by the application of two coats and catalysed clear lacquer (melamine) wherever it is indicated in the drawing/specified. The finish shall be a stain semi-gloss finish and shall be carried out as follows:-

The base shall be sand papered to the desired finish and coated with a colour tings to give it shade. This shade shall be sealed with a coat of spirit finish.

After the base, first coat of melamine shall be applied evenly by spray to give as even coat to the veneer surface.

After the first coat has fully dried, the surface shall be rubbed down in the direction of the veneer grain with very fine glass paper and left completely smooth and clean before the second coat is applied.

When the second coat of melamine is fully dry, the surface shall be rubbed down in the direction of veneer grain with very wire dipped in a petroleum based wax to give lubrication.

Twenty four hours after completion of this process the melamined veneer surface shall be finished by burnishing a soft cloth to an approved finish.

1.2.2 4.15 PLASTER OF PARIS PUNNING

Where specified the walls and ceilings are to be treated with plaster of Paris over cement plaster.

The particular brand of in gradients forming this special plaster and its composition to be used must be previously approved by the Engineer-in-Charge. The entire surface must be of a very smooth surface and minutest unevenness must be removed. Specially trained skilled labour with previous experience of this type work shall have to be employed for the purpose to achieve high grade finish.

1.2.3 4.16   GYPSUM PLASTER

1.2.4 Gypsum plasters; gypsum mill-mixer plaster, gypsum neat plaster, wood fibred plaster and gypsum gauging plaster are belowGypsum plaster, or plaster of Paris, is produced by heating gypsum to about 300°F (150 °C)

1.2.5 $CaSO_4 \cdot 2H_2O + Heat \rightarrow CaSO_4 \cdot \frac{1}{2}H_2O + 1\frac{1}{2} H_2O$ (released as steam)

1.2.6 When the dry plaster powder is mixed with water, it re-forms into gypsum. The setting of unmodified plaster starts about 10 minutes after mixing and is complete in about 45 minutes; but not fully set for 72 hours. If plaster or gypsum is heated above 392°F (200°C), anhydrite is formed, which will also re-form as gypsum if mixed with water

1.2.7 Applicable codes

IS 2542 Part-1 : Method test for gypsum plaster & concrete products

IS 2542   : Normal consistency of gypsum   concrete Part-1/ Sec-1

IS 2543 : Normal consistency of gypsum concrete Part-1/ Sec-2

IS 2542   : Setting time of plaster of paris Part-1/Sec-3

IS 2547 Part-1 : Specification for Gypsum plaster: Part-1 Excluding premixed light weight concrete

IS 2547 Part-2 : Specification of Gypsum plaster Part-2 remixed light weight concrete.

ASTM C28 :   Standard Specification for Gypsum Plaster

SP 21 : Summaries of Indian Standards for Building Materials

1.2.8 Gypsum Plaster work

1.2.9

1.2.10 Gypsum Gauging plaster for finished coat : A Calcined gypsum plaster designed to be mixed with lime putty for finished coat. Gypsum neat plaster : Calcined gypsum plaster mixed at mill with other ingredients to control working quality and setting time. Neat plaster is either fibred or un-fibred. The addition of aggregate is required on the job

1.2.11 Gypsum mill-mixed plaster : Calcined gypsum plaster, mixed at the mill with mineral aggregate, designed to functions a base coat to receive various finish coats. Other materials are not prohibited from being added to control setting time and other desirable working properties. Gypsum wood fibred plaster : A calcined gypsum plaster in which wood fibre is used as an aggregate

5.0   MISCELLANEOUS WORKS

5.1   STRUCTURAL STEEL WORK

5.1.1   This specification covers the fabrication and transportation to site and erection on prepared foundations and structural steel work consisting of beams, columns, vertical trusses, bracings, shear connections etc.

5.1.2   Fabrication, erection and approval of steel structures shall be in compliance with :

These General Specifications and IS : 800 - 1984

Drawings and supplementary drawings to be supplied to the contractors during execution of the work.

5.1.3   Providing primer coat for steel structures. Grouting of holding-down bolt pockets and below base plates where required.

5.1.4   In case of conflict between the Clauses mentioned here and the Indian Standards, those expressed in this specification shall govern.

5.2   Scope

5.2.1   The fabrication and erection of the steel work consists of accomplishing of all jobs here-in enumerated including providing all labour, tools and plant all materials and consumables such as welding electrodes, bolts and nuts, oxygen and acetylene gases, oils for cleaning etc. of approved quality as per relevant IS. The work shall be executed according to the drawings, specifications, relevant codes etc. in an expeditious and workman like manner, as detailed in the specifications and the relevant Indian Standard Codes and Standard Practice and to the complete satisfaction of the Engineer-in-Charge.

5.3   Fabrication Drawings

5.3.1   The contractor shall prepare all fabrication and erection drawings on the basis of design drawings supplied to him and submit the same in triplicate to the Engineer-in-Charge for review, Engineer-in-

Charge shall review and comment, if any, on the same. Such review, if any, by the Engineer-in-Charge, does not relieve the contractor of any of his required guarantees responsibilities. The contractor shall however be responsible to fabricate the structurals strictly conforming to specifications and reviewed drawings.

5.3.2   Fabrication drawings shall include the following :

-    Member sizes and details

-    Types and dimensions of welds and bolts

-    Shapes and sizes of edge preparation for welding

-    Details of shop and field joints included in assemblies.

Bill of material

-    Quality of structural steels, welding electrodes, bolts, nuts and washers etc. to be used.

-    Erection assemblies, identifying all transportable parts and sub-assemblies, associated with special erection instructions, if required.

-    Calculations where asked for, for approval.

5.3.3   Connections, splices etc. other details not specifically detailed in design drawings shall be suitably given on fabrication drawings considering normal detailing practices and developing full member strengths. Where asked for calculations for the merit shall also be submitted for approval.

5.3.4   Any alternate design or change in section is allowed when approved in writing by the Engineer-in-Charge.

5.3.5   However if any variation in the scheme is found necessary later, the contractor will be supplied with revised drawings. The contractor shall incorporate these changes in his drawings at no extra cost and resubmit for review.

5.3.6   Engineer-in-Charge review shall not absolve the contractor of his responsibility for the correctness of dimensions, adequacy of details and connections. One copy will be returned reviewed with or without comments to the contractor for necessary action. In the former case further three copies of amended drawings shall be submitted by the contractor for final review.

5.3.7   The contractor shall supply three prints each of the final reviewed drawings to the Engineer-in-Charge within a week since final review, at no extra cost for reference and records.

5.3.8   The Engineer-in-Charge will verify the correct interpretation of their requirements.

5.3.9   If any modification is made in the design drawing during the course of execution of the job, revised design drawings will be issued to the contractor. Further changes arising out of these shall be incorporated by the contractor in the fabrication drawings already prepared at no extra cost and the revised fabrication drawings shall be duly got reviewed as per the above Clauses.

5.4 **Materials**

5.4.1 Rolled Sections

The following grades of steel shall be used for steel structures :

Structural steel will generally be of standard quality conforming to IS: 226. Whenever welded construction is specified plates of more than 20 mm thickness will generally conform to IS: 2062.

5.4.2 Welding Materials

Welding electrodes shall conform to IS: 814.

Approval of welding procedures shall be as per IS: 823.

5.4.3 Bolts, Nuts & Washers

Bolts and nuts shall be as per IS: 1367 and tested as per IS:1608. It shall have a minimum tensile strength of 44 Kg/mm2 and minimum elongation of 23% on a gauge length of 5.65 (A- Original cross sectional area of the gauge length). Washers shall be as per IS: 2016.

5.4.4   All materials shall conform to their respective specifications. The use of equivalent or higher grade or alternate materials will be considered only in very special cases subject to the approval of the Engineer-in-Charge in writing.

5.4.5   Receipt & Storing of Materials

Steel materials supplied by the contractor must be marked for identification and each lot should be accompanied by manufacturer's quality certificate, conforming chemical analysis and mechanical characteristics.

All steel parts furnished by supplier shall be checked, sorted out, straightened, and arranged by grades and qualities in stores.

Structurals with surface defects such as pitting, cracks, laminations etc. shall be rejected if the defects exceed the allowable tolerances specified in relevant standards or as directed by the Engineer-in-Charge.

Welding wire and electrodes shall be stored separately by qualities and lots inside a dry and enclosed room, in compliance with IS: 816 - 1969 and as per instructions given by the Engineer-in-Charge. Electrodes shall be perfectly dry and drawn from an electrode even, if required.

Checking of quality bolts of any kind as well as storage of same shall be made conforming to relevant standards.

Each lot of electrodes, bolts, nuts, etc. shall be accompanied by manufacturer's test certificate.

The contractor may use alternative materials as compared to design specification only with the written approval of the Engineer-in-Charge.

5.4.6   Material Tests

The contractor shall be required to produce manufacturer's quality certificates for the materials supplied by the contractor. Notwithstanding the manufacturer's certificates, the Engineer-in-Charge may ask for testing of materials in approved test houses. The test results shall satisfy the requirements of the relevant Indian Standards.

Whenever quality certificates are missing or incomplete or when material quality differs from standard specifications the contractor shall conduct all appropriate tests as directed by the Engineer-in-Charge at no extra cost.

Materials for which test certificates are not available or for which test results do not tally with relevant standard specifications, shall not be used.

5.5   **Fabrication**

Fabrication shall be in accordance with IS: 800 Section V in addition to the following :

Fabrication shall be done as per approved fabrication drawings adhering strictly to work points and work lines on the same. The connections shall be welded or bolted as per design drawings. Work shall also include fabricating built up sections.

Any defective material used shall be replaced by the contractor at his own expense, care being taken to prevent any damage to the structure during removal.

All the fabricated and delivered items shall be suitably packed to be protected from any damage during transportation and handling. Any damage caused at any time shall be made good by the Contractor at his own cost.

Any faulty fabrication pointed out at any stage of work shall be made good by the contractor at his own cost.

5.5.1   Preparation of Materials

Prior to release for fabrication, all rolled sections warped beyond allowable limit shall be pressed or rolled straight and freed from twists, taking care that an uniform pressure is applied.

Minor warping, corrugations etc. in rolled sections shall be rectified by cold working.

The sections shall be straightened by hot working where the Engineer-in-Charge so direct and shall cooled slowly after straightening.

Warped members like plates and flats may be used as such only if wave like deformation does not exceed L/1000 but limited to 10 mm (L-Length).

Surface of members that are to be jointed by lap or fillet welding or bolting shall be even so that there is no gap between overlapping surfaces.

5.5.2   Marking

Marking of members shall be made on horizontal pads, of an appropriate racks or supports in order to ensure horizontal and straight placement of such members.

Marking accuracy shall be atleast + 1 mm.

5.5.3   Cutting

Members shall be cut mechanically (by saw or shear or by oxyacetylene flame).

All sharp, rough, or broken edges, and all edges of joints which are subjected to tensile or oscillating stresses, shall be ground.

No electric metal arc cutting shall be allowed.

All edges cut by oxyacetylene process shall be cleaned of impurities prior to assembly.

Cutting tolerances shall be as follows :

a) For members connected at both ends + 1 mm.
b) Elsewhere + 3 mm.

The edge preparation for welding of members more than 12 mm thick shall be done by flame cutting and grinding. Cut faces shall not have cracks or be rough.

Edge preparation shall be as per IS : 823 - 1964.

5.5.4   Drilling

Bolts holes shall be drilled.

Drilling shall be made to the diameter specified in drawings.

No enlarging of holes filling, by mandrolling or oxyacetylene flame shall be allowed.

Allowed variations for holes (out-of-roundness, eccentricity, plumb-line deviation) shall be as per IS:800.

-    Maximum deviation for spacing of two holes on the same axis shall be + 1 mm.

- Two perpendicular diameters of any oval hole shall not differ by more than 1 mm.

Drilling faults in holes may be rectified by reaming the holes to the next upper diameter, provided that spacing of new hole centres and distance of hole centres to the edges of members are not less than allowed and that the increase of hole diameter does not impair the structural strength. Hole reaming shall be allowed if the number of faulty holes does not exceed 15% of the total number of holes for one joint.

5.5.6   Preparation of Members for Welding

Assembly of structural members shall be made with proper jigs and fixtures to ensure correct positioning of members (angles, axes nodes etc.)

Sharp edges, rust of cut edges, notches, irregularities and fissures due to faulty cutting shall be chipped or ground or filled over the length of the affected area, deep enough to remove faults completely.

Edge preparation for welding shall be carefully and accurately made so as to facilitate a good joint.

Generally no special edge preparation shall be required for members under 8 mm thick.

Edge preparation (bevelling) denotes cutting of the same so as to result in V, X K or U seam shapes as per IS: 823.

The members to be assembled shall be clean and dry on the welding edges. Under no circumstances shall wet, greasy, rust or dirt covered parts be assembled. Joints shall be kept free from any foreign matter likely to get in to the gaps between members to be welded.

Before assembly the edges to be welded as well as adjacent areas extending for atleast 20 mm shall be cleaned (until metallic polish is achieved).

When assembling members, proper care shall be taken of welding shrinkage and distortions, as the drawing dimensions cover finished dimensions of the structure.

The elements shall be got checked and approved by the Engineer-in-Charge or their authorised representative before assembly.

The permissible tolerances for assembly of members preparatory to welding shall be as per IS: 823-1964.

After the assemble has been checked, temporary tack welding in position shall be done by electric welding, keeping in view finished dimensions of the structure.

5.5.7   Welding procedures

Welding shall be carried out only by fully trained and experienced welders as tested and approved by the Engineer-in-Charge. Any test carried out either by the Engineer-in-Charge of their representative or the inspectors shall constitute a right by them for such tests and the cost involved thereon shall be borne by the contractor himself.

Qualification tests for welders as well as tests for approval of electrodes will be carried out as per IS: 823. The nature of test for performance qualification of welders shall be commensurate with the quality of welding required on this job as judged by the Engineer-in-Charge.

The steel structures shall be automatically, semi-automatically or manually welded.

Welding shall begin only after the checks mentioned in Clause 5.1 to 5.6 have been carried out.

The welder shall mark with his identification mark on each element welded by him.

When welding is carried out in open air, steps shall be taken to protect the face of welding against wind or rain. The electrodes, wire and parts being welded shall be dry.

Before beginning the welding operation, each joint shall be checked to ensure that the parts to be welded are clean and root gaps provided as per IS: 823.

For continuing the welding of seems discontinued due to some reason, the end of the discontinued seem shall be melted in order to obtain a good continuity. Before resuming the welding operation, the groove as well as the adjacent parts shall be well cleaned for a length of approx. 50 mm.

For single butt welds (in V, 1/2 V or U) and double butt welds (in K, double U etc.) the rewelding of the root is mandatory but only the metal deposit on the root has been cleaned by back gouging or chipping.

The welding seams shall be left to cool slowly. The contractor shall not be allowed to cool the welds quickly by any other method.

For multi-layer welding, before welding the following layer, the formerly welded layer shall be cleaned metal bright by light chipping and wire brushing. Backing strips shall not be allowed.

The order and method of welding shall be so that -

-    No unacceptable deformation appears in the welded parts.

-    Due margin is provided to compensate for contraction due to welding in order to avoid any high permanent stresses.

The defects in welds must be rectified according to IS: 823 and as per instruction of Engineer-in-Charge.

5.5.8   Weld Inspection

The weld seams shall satisfy the following :

-    shall correspond to design shapes and dimensions.

-    shall not have any defects such as cracks, incomplete penetration and fusion, under-cuts, rough surfaces, burns, blow holes and porosity etc. beyond permissible limits.

During the welding operation and approval of finished elements, inspections and tests shall be made as shown in annexure-B.

The mechanical characteristics of the welded joints shall be as in IS: 823.

5.5.9   Preparation of Members for Bolting

The members shall be assembled for bolting with proper jigs and fixtures to sustain the assemblies without deformation and bending.

Before assembly, all sharp edges, shavings, rust dirt, etc. shall be removed.

Before assembly, the contacting surfaces of the members shall be cleaned and given a coat of primer as per IS: 2074.

The members which are bolt assembled shall be set according to drawings and temporarily fastened with erection bolts (minimum 4 pieces) to check the coaxiality of the holes.

The members shall be finally bolted after the deviations have been corrected, after which there shall not be gaps.

Before assembly, the members shall be checked and got approved by the Engineer-in-Charge.

The difference in thickness of the sections that are butt assembled shall not be more than 3% or maximum 0.8 mm whichever is less. If the difference is larger, it shall be corrected by grinding or filling.

Reaming of holes to final diameter or cleaning of these shall be done only after the parts have been check assembled.

As each hole is finished to final dimensions (reamed if necessary) it shall be set and bolted up. Erection bolts shall not be removed before other bolts are set.

5.5.10   Bolting up

Final bolting of the members shall be done after the defects have been rectified and approval of joints obtained.

The bolts shall be tightened starting from the centre of joint towards the edge.

5.5.11   Planing of Ends

Planing of ends of members like column ends shall be done by grinding when so specified in the design.

Planing of butt welded members shall be done after these have been assembled, the spare edges shall be removed with grinding machines or files.

The following tolerances shall be permitted on member that have been planed.

- On the length of the member having both ends planed, maximum + 2 mm with respect to design.

- Level differences of planed surfaces, maximum 0.3 mm.

- Deviation between planed surface and member's axis maximum 1/1500.

5.5.12   Holes for Field Joints

Holes for field joints shall be drilled in the shop to final diameters and tested in the shop, with trial assemblies.

When three-dimensional assembly is not possible in the shop, the holes for field joints may be drilled in shop and reamed on site after erection, on approval by the Engineer-in-Charge.

For bolted steel structures, trial assembly in shop is mandatory.

The tolerance for spacing of holes shall be + 1 mm.

5.5.13   Tolerances

All tolerances regarding dimensions, geometrical shapes and sections of steel structures, shall be as per Annexure B, if not specified in the drawing.

5.5.14   Marking for Identification

All elements and members prior to despatch for erection shall be shop marked.

The members shall be visibly marked with a weather proof light coloured paint. The size and thickness of the numbers shall be chosen as to facilitate the indentification of members.

For the small members that are delivered in bundles or crates, the required marking shall be done on small metal tags securely tied to the bundle, while the crates shall be marked directly.

Each bundle or crate shall be packed with members for one and the same assembly; in the same bundle or crate, general utility members such as bolts, quests etc. may be packed.

All bill of materials showing weight, quality and dimension of contents shall be placed in the crates.

The members shall be marked with a durable paint, in a visible location, preferably at one end of the member so that these may be easily checked during storage and erection.

All members shall be marked in the shop before inspection and acceptance.

When the member is being painted, the marking area shall not be painted but bordered with white paint.

The marking and job symbol shall be registered in all shop delivery documents (transportation, for erection etc.)

5.5.15   Shop Test Pre-assembly

For steel structures that have the same type of welding the shop test pre-assembly shall be performed on one out of every 10 members minimum.

For bolted steel structures, shop test pre-assembly is mandatory for all elements as well as for the entire structure in conformity with Clause 5.12.

5.6 Shop Inspection and Approval

5.6.1 General

The Engineer-in-Charge or their representative shall have free access at all responsible times to the contractors fabrication shop and shall be afforded all reasonable facilities for satisfying himself that the fabrication is being undertaken in accordance with drawings and specifications.

Technical approval of the steel structure in the shop by the Engineer-in-Charge is mandatory.

The contractor shall not limit the number and kinds of tests, final as well as intermediate once, or extra tests required by the Engineer-in-Charge.

The contractor shall furnish necessary tools, gauges, instruments etc. and technical non-technical personnel for shop tests by the Engineer-in-Charge, free of cost.

5.6.2 Shop Acceptance

The Engineer-in-Charge shall inspect and approve at the following stages :

The following approvals may given in shop :

- Intermediate approvals of work that cannot be inspected later.

- Partial approvals

- Final approvals

Intermediate approval of work shall be given when a part of the work is preformed later :

- Cannot be inspected later

- Inspection would be difficult to perform and results would not be satisfactory.

Partial approval in the shop is given on members and assemblies of steel structures before the primer coat is applied and includes :

- Approval of materials
- Approval of field joints
- Approval of parts with planed surfaces
- Test erection

- Approval of members
- Approval of markings
- Inspections and approvals of special features, like Rollers, loading platform mechanism etc.

During the partial approval, intermediate approvals as well as all former approvals, shall be taken in to consideration.

5.6.3 Final approval in the Shop

The final approval refers to all elements and assemblies of the steel structures, with shop primer coat, ready for delivery from shop to be loaded for transportation, or stored.

The final approval comprises of :

- Partial approvals
- Approval of shop primer coat
- Approval of mode of loading and transport
- Approval of storage (for materials stored)

5.7 Painting and Delivery

5.7.1 Preparation of parts for shop painting

Painting shall consist of providing one coat of red oxide zinc chromate primer to steel members before despatch from shop.

Primer coat shall not be applied unless :

- Surface have been wire brushed, cleaned of dust, oil, rust etc.

- Erection gaps between members, spots that cannot be painted or where moisture or other aggressive agents may penetrate, have been filled with an approved type of oil and putty.

- The surface to be painted are completely dry.

- The parts where water of aggressive agents may collect (during transportation, storage, erection and operation) are filled with putty and provided with holes for drainage of water.

- Members and parts have been inspected and accepted

- Welds have been accepted.

The following are not to be painted or protected by any other product :

- Surface which are in the vicinity of joints to be welded at site.

- Surfaces bearing markings

- Other surfaces indicated in the design.

The following shall be given a coat of hot oil or any approved resistant lubricant only.

- Planed surfaces
- Holes for links

The surfaces that are to be embedded or in contact with the concrete shall be given a cost of cement wash.

The surfaces which are in contact with the ground, gravel or brick work and subject to moisture, shall be given bituminous coat.

The other surfaces shall be given a primer coating.

Special attention shall be given to locations not easily accessible, where water can collect and which after assembly and erection cannot be inspected, painted and maintained. Holes shall be provided for water drainage and in accessible box type sections shall be hermetically sealed by welds.

If specified elsewhere, in the schedule of quantities, the contractor shall paint further coats of red-oxide after erection and placing in position of the steel structures.

5.7.2   Packing, transportation, delivery

After final shop acceptance and marking, the item shall be packed and loaded for transportation.

Packing must be adequate to protect item against warping during loading and unloading.

Proper lifting devices shall be used for loading, in order to protect items against warping.

Slender projecting parts shall be braced with additional steel bars, before loading, for protection against warping during transportation.

Loading and transportation shall be done in compliance with transportation rules.

If certain parts cannot be transported in the lengths stipulated in the design, the position and type of additional splice joints shall be approved by the Engineer-in-Charge.

Items must be carefully loaded on platforms of transportation means to prevent warping, bending or falling during transportation.

The small parts such as fish-plates, quests etc. shall be securely tied with wire to their respective parts.

Bolts, nuts and washers shall be packed and transported in crates.

The parts shall be delivered in the order stipulated by the Engineer-in-Charge and shall be accompanied by document showing:

- Quality and quantity of structure or members
- Position of member in the structure
- Particulars of structure
- Identification number job symbol.

5.8 Field Erection

5.8.1 The erection work shall be permitted only after the foundation or other structure over which the steel work will be erected is approved and is ready for erection.

5.8.2 The contractor shall satisfy himself about the levels, alignment etc. for the foundations well in advance, before starting the erection. Minor chipping etc. shall be carried out by the contractor on his expense.

5.8.3 Any faulty erection done by the contractor shall be made good at his own cost.

5.8.4 Approval by the Engineer-in-Charge or their representatives at any stage of work does not relieve the contractor of any of his required guarantees of the contract.

5.8.5 Storage and preparation of parts prior to erection

The storage place for steel parts shall be prepared in advance and got approved by the Engineer-in-Charge before the steel structures start arriving from the shop.

A platform shall be provided by the Contractor near the erection site for preliminary erection work.

The contractor shall make the following verifications upon receipt of material at site.

- for quality certificates regarding materials and workmanship according to these general specifications and drawings.
- Whether parts received are complete without defects due to transportation, loading and unloading and defects, if any, are well within the admissible limit.

For the above work sufficient space must be allotted in the storage area.

Steps shall be taken to prevent warping of items during unloading.

The parts shall be unloaded, stored and stored so as to be easily identified.

The parts shall be stored according to construction symbol and markings so that these may be taken out in order or erection.

The parts shall be at least 150 mm clear from ground on wooden or steel blocks for protection against direct contact with ground and to permit drainage of water.

If rectification of members like straightening etc. are required, these shall be done in a special place allotted which shall be adequately equipped.

The parts shall be clean when delivered for erection.

5.8.6   Erection & Tolerances

Erection in general shall be carried out as required and approved by the Engineer-in-Charge.

Positioning and levelling of the structure, alignment and plumbing of the stanchion and fixing every member of the structure shall be in accordance with the relevant drawings and to the complete satisfaction of the Engineer-in-Charge.

The following checks and inspection shall be carried out before during and after erection.

- damage during transportation
- accuracy of alignment of structures

- erection according to drawings and specifications

- progress and workmanship.

In case there be any deviations regarding positions of foundations or anchor bolts, which would lead to erection deviations, the Engineer-in-Charge shall be informed immediately. Minor rectifications in foundations, orientation of bolts holes etc. shall be carried out as part of the work, at no extra cost.

The various parts of the steel structure shall be so erected so to ensure stability against inherent weight, wind and erection stresses.

The structure shall be anchored and final erection joints completed after plan and elevation positions of the structural members have been verified with corresponding drawings and approved by the Engineer-in-Charge.

The bolted joints shall be tightened so that the entire surface of the bolt heads and nuts shall rest on the member. For parts with sloping surfaces tapered washers shall be used.

5.9   Final acceptance and handing over the structure

5.9.1   At acceptance, the contractor shall submit the following documents :

- Shop and erection drawings - either in tracings or reproducible.

- 4 copies of each of the following :

  \- shop acceptance documents

  \- quality certificate for structurals, plates, etc. (electrodes, welding wire, bolts, nuts, washers etc.)

  \- List of certified welders who worked on erection of structures.

  \- acceptance and intermediate control procedure of erection operations.

5.9.2 Approval by the Engineer-in-Charge at any stage of work does not relieve the contractor of any of his required guarantees of the contract.

5.10 Method of Payments

5.10.1 Payment for steel work shall be made on basis of admissible weight of the structure accepted, the weight being determined as described in such Clause 9.10.2 below :

The rate for supply, fabrication and erection, shall include cost of all handling and transportation to Owner's store/site o work where supply and fabrication only are involved, trimming, straightening, edge preparation, preparation and getting reviewed of fabrication drawings, and providing one or more coat of Red-oxide zinc chromate primer as specified in the schedule of quantity.

In the case, Owner supplies materials the rate shall include cost of steel materials taking delivery of the materials, from owner's store all handling and rehandling, loading and unloading, transport to site or work, returning of surplus materials to owner's stores etc. complete as well as the cost of all handling and transport, scaffolding, temporary supports, tools and tackles, touching up primer coat, grouting etc.

5.10.2   The actual lengths installed shall be measured and the weight of structural material/plate shall be calculated wherever necessary on the basis of IS handbook.   If sections are different from IS section, then manufacturers handbook shall be adopted. No allowance in weights shall be made for rolling tolerance.

5.10.3   Sections built out of plates, structural shall be paid on the actual weight incorporated except for gussets which will be paid on the weight of the smallest rectangle enclosing the shape.   No deductions shall be made for skew cuts in rolled steel sections.

5.10.4   Welds, bolts, nuts, washers, etc. shall not be measured. Rate for structural steel work shall be deemed to include the same.

5.10.5   No other payment either for temporary works connected with this contract or for any other item such as welds, shims, pacing plates etc. shall be made. Such item shall be deemed to have been allowed for in the rate quoted for steel work.

5.11   Grouting of Pockets

5.11.1   Grouting of pockets and under base plates will be done only after the steel work has been levelled and plumbed and the bases of stranchions are supported by steel shims. The space below the base plate and pockets shall be thoroughly cleaned.

5.11.2   The mortar used for grouting shall not be leaner than 1:2 (1 cement : 2 sand) (grade 300 in case of concrete) and shall be mixed to the minimum consistency required. It shall be poured under suitable head and tamped until the space has been completely filled.

5.12   Tolerances allowed in the erection of plant building without cranes

The maximum tolerances for line and level of the steel work shall be + 3.00 mm on any part of the structure. The structure shall not be out of plumb more than 3.5 mm on each 10 M. section of height and not more than 7.0 mm per 30 M. section.

These tolerances shall apply to all parts of the structure unless the drawings issued for erection purposes state otherwise.

5.13   **Rolling Shutters**

Rolling shutters shall be in extruded MS sections, of approved make, type and finish. These shutters shall be complete with locking arrangements, hoods, guides, pulling devices, springs and other accessories. Wherever specified, mechanical device shall be fixed for easy operation of the shutters. (Rolling shutters as per CPWD Specification Volume -1, Clause No. 10.8, Page No. 564).

5.14   Steel Door / Window

Hot rolled steel sections for fabrication of steel doors, windows, ventilators and fixed lights shall conform to IS : 7452. Shapes weights and designations of hot rolled sections shall be as per IS : 7452. Appendix 'D' of Chapter 10 (CPWD specification 2000) indicates the purpose or the situation where the sections are normally used. Tolerance in thickness of the sections shall be +0.2mm including provisions of lugs or anchoring strips, both for RCC and Brickwork backing.

5.14.1   The steel doors and windows shall be according to the specified sizes and design. The size of doors and windows shall be calculated, so as to allow 1.25 cm clearance on all the four sides of opening to allow for easy fitting of doors windows and ventilators into opening. The actual sizes of doors, windows and ventilators shall not vary by more than +1.5mm from those given in the drawing.

5.14.2   Fabrication

5.14.2.1 Frames

Both the fixed and opening frames shall be made of sections which have been cut to length and metered. The corner of fixed and opening frames shall be welded to form a solid fused welded joint conforming to the requirements given below. All frames shall be square and flat. The process of welding adopted shall be flash butt welding. The section for glazing shall be tennoned and riveted into the frames and where they intersect the vertical tie shall be broached and horizontal tee threads through it, and the intersection closed by hydraulic pressure.

Requirements of welded joints

i)   Visual Inspection Test

When two opposite corners of the frame are cut, paint removed and inspected, the joint shall conform to the following:

a) Welds should have been made all along the place of meeting the members and tack welding shall not be permitted.
b) Welds should have been properly grounded and
c) Complete cross section of the corner shall be checked up to see that the joint is completely solid and there are no cavaties visible.

ii)   Micro and Macro Examinations

From the two opposite corners obtained for visual test, the flanges of the sections shall be cut with the help of the sections shall be cut with the help of a saw. The cut surface of the remaining portions shall be polished, etched and examined. The polished and etched faces of the weld and the base metal shall be free from cracks and reasonably free from under cutting, overlaps, gross porosity and entrapped slag.

iii)   Fillet Weld Test

The fillet weld in the remaining portion of the joint shall be fractured by hammering. The fractured surfaces shall be free from slag inclusion porosity, crack penetration defects and fusion defects.

5.14.2.2 Door

The hinges shall be of 50mm projecting type, Non projecting type hinges may also be used if approved by the Engineer-in-Charge. The hinge pin shall be of electro-galvanized steel or aluminium alloy of suitable thickness and size. Door handles shall be approved by the Engineer-in-Charge. A suitable latch lock for door openable both from inside and outside shall be provided.

In the case of double doors, the first closing leaf shall be the left hand leaf locking at the door from the push side. The first closing shutter shall have a concealed steel bolt at top and bottom. The bolts shall be so constructed as not to work loose or drop by its own weight.

Single and double shutter door may be provided with a three way bolting device. Where the device is provided in the case of double shutters, concealed brass or steel bolts shall not be provided.

5.14.2.3 Windows

a) For fixed windows, the frames shall be fabricated as per relevant CPWD specification 2000.
b) Side hung windows.

For fixing steel hinges, slots shall be cut in the fixed frame and hinges inserted inside and welded to the frame at the back . The hinges shall be of projecting type with thickness not less than 3.15 mm and length not less than 65mm and width not more than 25mm. Non-projecting type hinges may also be allowed if approved by the Engineer-in-Charge. The diameter of hinge pins shall not be less than 6mm. The hinge pin and washer shall be of galvanised steel or aluminium alloy of suitable thickness.

For fixing hinges to inside frame, the method described above may be adopted but the weld shall be cleaned, or the holes made in the inside frame and hinge revetted.

The handle of side hung shutters shall be pressed brass, cast brass, aluminium or steel protected against rusting and shall be mounted on a steel plate. Thickness of handle shall not be less than 3 mm incase of steel or brass and 3.5mm in ase of aluminium. The handle plate shall be welded, screwed and/or revetted to the opening frame in such a manner that it should be fixed before the shutter is glazed and should not be easily removable after glazing.

The handle shall have a two point nose which shall engage with a brass or aluminium alloy striking plate on the fixed frame in a slightly opened position as well as closed position. The boss of handle shall incorporate a friction device to prevent the handle shall incorporate a friction device to prevent the handle from dropping under its own weight and the assembly shall be so designed that the rotation of the handle may not cause it to unscrew from the pin.

The height of the handle plate in each type of standard windows will be as specified. Otherwise it shall be at a height of 3/8 of the height of shutter, from its bottom. The strike plate shall be so designed and fixed in such a position in relation to the handle that with the later bearing against its stop, there shall be adequately tight fit between the casement and outer frames.

In case where no friction type hinges are provided, the window shall be fitted with peg stays which shall be either of black oxidised steel, pressed or cast brass or as specified, 300mm long with steel peg and locking brackets. The pegs stay shall have three holes to open the side hung casement in three different angles. The peg stay shall be of minimum thickness 2mm in case of brass or aluminium and 1.25mm in case of steel. Where specified friction hinges shall be provided. Side hung shutters fitted with friction hinges shall not be provided with a peg stay.

If specified, side hung shutters maybe fitted with an internal removable fly proof screen in a 1.25mm thick sheet steel frame to the outer frame of the shutter by brass turn buckless at the jambs, and brass tuds at the sill to allow the screen being readily removed. The windows with removable fly proof screen shall be fitted with a through – the – screen level operator at the sill level to permit the operation of the shutter through an angle of 90o without having to remove the fly proof screen. The lever shall permit keeping the shutter open in minium three different position.

5.14.2.4 Ventilators

a)   Top Hung Ventilators

The steel butt hinges for top hung ventilators shall be riveted to the fixed frame or welded to it at the back after cutting a slot in it. Hinges to the opening frame shall be revetted or welded.

Top hung ventilators shall be provided with a peg stay with three holes which when closed shall be held tightly by the locking bracket. The locking bracket shall either be fitted to the fixed frames or to the window.

b)   Centre Hung Ventilators

Central hung ventilators shall be hung on two pairs of brass or aluminium cup pivots as specified, riveted to the inner and outer frames of ventilators to permit the ventilator shutter to swing to an angle of approx. 85o. The opening portion of the ventilators shall be so balanced that it remains open at any desired angle under normal weather conditions.

5.15   Glazing

5.15.1   Specifications as described shall apply. The glass panes shall have square corners and straight edges. The glass panes shall be so cut that it fits slightly loose in the frames. In doors, windows, clerestory windows of bath, WC and lavatories frosted glass panes shall be used.

5.15.2   Glazing shall be provided on the outside of the frame unless otherwise specified. Putty of approved make conforming to IS:419 shall be used for fixing glass panes. Putty shall be applied between glass panes and glazing bars. Putty shall then be applied over the glass pane, which shall stop 2 to 3 mm from the sight line of the back rebate to enable the painting to be done upto the sight line to seal the edge of the putty to the glass. The oozed out putty shall be cleaned and from putty cut to straight line. Quantity of putty shall not be less than 185 gm/metre of glass perimeter. Putty shall be painted within 2 to 3 weeks, after glazing is fixed to avoid its cracking.

Note: Putty may be prepared by mixing one part of the white lead with three parts of finely powdered chalk and then adding boiled linseed oil to the mixture to from a stiff paste and adding varnish to the past at the rate of 1 litre of varnish to the paste at the rate of 1 litre of varnish to 18 kg of paste.

5.15.3   Minimum Six glazing clips to be provided per glass pane of all types. In case of doors, windows and ventilators without horizontal glazing bars, the glazing clips may be spaced, according to the slots, in the vertical members provided the spacing does not exceed 30cm otherwise the spacing shall be 30cm.

Note: Where large size glass panes are required to be used or where the door window is located in heavily exposed situation, holes for glazing clips have to be drilled prior to fabrication and cannot be done at any later stages.

5.15.4   Where specially stipulated, fixing of glass panes may be done with metal or wooden beading instead of mere putty. Where beading are proposed to be used, the manufacturers shall be intimated in advance to drill holes for hard screws. Usually beads shall be fixed with screws spaced not more than 10 cm from each corner and the intermediate not more than 20 cm apart. When glass panes are fixed with wooden or metal beading having mitred joints, a thin layer of putty shall be applied between glass panes and sash bars and also between glass panes and the beading.

Where metal beading is specified extra payment shall be made on this account.

5.16   T-Iron / Angle Iron Doors Frames

T-iron / Angle Iron doors, frames shall be manufactured from uniform mild steel tee section / angle section. The steel shall be of approved grade. The frames shall be got fabricated in approved workshop as approved by the Engineer-in-Charge. Provisions of lugs or anchoring strips, both for RCC and Brickwork backing.

5.16.1   The sizes of door frames shall be as per drawing or as decided by the Engineer-in-Charge. MS tie bar of 10mm dia shall be welded at bottom of the frame. The size of doors shall be calculated so as to allow 12.5mm clarance on all sizes to allow an easy fittings in opening. The actual size of doors, windows and ventilators shall not vary by more than +/- 2mm than those shown in the drawing.

The size of T-section / Angle section used for manufacture of doors, windows and ventilators shall not be less than those specified in IS:1038.

Unless otherwise directed by the Engineer-in-Charge.

5.16.2 Fabrications

The frame shall be constructed in section which has been cut to length and metered. The corners of the frames shall be butt welded to form a true and right angle. All frames shall be square and flat.

The T-section / Angle section shall be mitre joined and continuously butt welded all along.

5.16.3 Fittings

Requisite number of holes shall be made in the frame for fixing of fittings. Detailed arrangements of fixing fittings shall be as shown. All fittings shall be fillet welded to T-iron / Angle iron frame all along the periphery of contact.

Butt hinges shall be fixed to the frame as below:

i) MS flat of size 100mm x 25mm x 6mm will be welded with fillet weld all along the periphery of contact on the rear side of the web of T-iron / Angle iron to receive the hinges. Requisite number of holes shall be made in T-iron / Angle iron frame and MS flat for fixing of hinges with counter sunk steel screws as shown.

ii)   An alternate method of fixing butt hinges can be adopted by fillet welding the hinge to the T-iron / Angle iron frame on three sides. No welding shall be done along the hinge pin to allow free movement of butt hinges as shown.

5.16.4   Fixing Procedure

Fixing procedure for T-iron / Angle iron doors, windows and ventilator frames in masonry opening shall be as shown in the drawing.

5.17 Gypsum Board False Ceiling

5.17.1 Products

A   **Manufactures**

Manufacturer: Indian Gypsum Ltd. or approved equivalent, for gypsum board and all accessories, suspension systems, finishing coats etc.

5.17.1.1 **Gypsum Board**

A   General : Provide gypsum board of types indicated in maximum lengths available to minimise end to end joints.

1.   Thickness: Gypsum board in thickness indicated, in 12.5mm thickness to comply with ASTMC 840 for application system and support spacing indicated.

5.17.1.2 **Trim Accessories**

A   **Cornerbead and Edge Trim for Interior Installation:** G.I. corner beads, edge trim and control joints which comply with ASTM C 1047 to be provided as per requirements and drawings.

5.17.1.3 Gypsum Board Joint treatment Materials

A General : Materials complying with ASTM C 475, ASTM C 840 and recommendations of manufacturer or both gypsum board and joint treatment materials for the application indicated.

B Joint Tape : Paper reinforcing tape.

C Setting Type Joint Compounds : Factory – Pre-packaged, job-mixed, chemical hardening powder products formulated for uses indicated.

5.17.2 Execution

5.17.2.1 Installation of G.I. Steel Framing

A Hangers to be secured to structural support by connecting directly to structure where possible, otherwise to be connected to cast-in concrete inserts or other anchorage devices fasteners as required.

B Hangers are not to be connected or suspended from steel framing from ducts, pipes or conduits.

C To keep hangers and braces 50mm clear of ducts, pipes and conduits.

D To install suspended steel framing components in size and at spacing indicated but not less than that required by referenced steel framing installation standard.

E   Installation Tolerances : To install steel framing components for suspended ceilings so that cross furring members or grid suspension members are level to within 3mm in 4M as measured both lengthwise on each member and transversely between parallel members.

F   Wire-tie or clip furring members to main runners and to other structural supports as indicated.

G   Grid Suspension System: Perimeter wall track or angle not to touch where grid suspension system meets vertical surfaces, suspended from structure above. Mechanically join main beam and cross furring members to each other and butt-cut to fit into wall track. Wall track to be free moving from vertical wall surfaces.

H   For exterior soffits to provide cross-bracing and additional framing indicated or required to resist wind uplift.

I   For isolated ceilings to hold perimeter 12mm away from adjacent partitions to prevent flanking. The openings are to be sealed with compressible weather strip type edge seal to allow vertical movement.

5.17.2.2 Application and finishing of Gypsum Board, General

A   Gypsum Board Application and Finishing Standard: shall be to comply with ASTM C 840.

B   To install sound attenuation blankets where indicated, prior to gypsum board unless readily installed after board has been installed.

C   To locate exposed end-butt joints as far from centre of walls and ceilings as possible, and stagger not less than 600mm in alternate courses of board.

D   To install ceiling boards across traming in the manner which minimize the number of end-butt joints, and which avoids end joints central area of each ceiling. Stagger end joints at least 600mm.

E   To attach gypsum board to supplementary framing and blocking provided for additional support at openings and cutouts

PREAMBLE TO BOQ FOR FINISHING WORKS -

PREAMBLE TO SPECIFICATIONS

GENERAL

The conditions of contract and the drawings shall be read in conjunction with the specifications and matters referred to, shown or described in one are not necessarily repeated in the other. These specifications are comprehensive but may exceed the requirements of this project. Any ambiguity between the General Specifications, the Bill of quantities and contract drawings, shall be referred to the Project Manager for clarification not later than 10 days before the date fixed for delivery of Tenders. Any ambiguity may be referred to the Project Manager after signing of the contract and Project Manager shall give a ruling which shall prevail. No claim for additional cost due to above, however, will be entertained.

Notwithstanding the sub-division of the specification into various headings, every part of it is to be deemed supplementary to every other part and is to be read with it, so far as it may be practicable so to do, or when the context so admits.

In this contract, reference is made to the Indian Standards or CPWD specification as approved by Project Manager and these references shall be deemed to include the latest editions or issue of standards, specifications or By-Law including all revisions upto the date of invitation of Tenders. The contractor shall ensure that all materials and workmanship in so far as they apply to this contract shall comply in every specifications or any other equivalent or specification approved by the Project Manager.

The Contractor shall keep at site copies of all relevant standards and codes of practice referred in these specifications throughout the period of contract. These shall be the latest editions and shall include all revisions/addendums thereof.

Approved Manufacturers: Names of approved manufacturers are given in the specifications.

Reference in the specifications to approved manufacturers shall be construed as establishing a standard of quality and not as limiting competition.

The Contractor shall include in his prices for supplying the item or materials from the approved manufacturers listed or equal and approved.

All items or materials shall be delivered to the site in the manufacturers original unopened containers with the manufacturers brand and name clearly marked on.

All items or materials shall be assembled, mixed, fixed, applied or otherwise incorporated in the works in accordance with the printed instructions of the manufacturer of the item or materials.

Date of construction to be written on all respective items for monitoring curing.

Contractor shall submit shop drawings for the following but not limited to:
- Stone Cladding
- Doors
- Built in Furniture's
- Decorative Feature Elements

2   CONTRACT SHALL SUBMIT SAMPLE OF THE ALL FINISHING MATERIALS AS DIRECTED BY THE PROJECT MANAGER

3   BEFORE EXECUTION.

Contractor shall also prepare / fabricate Mock-ups for specialized works / finishing item for the following item but not limited to as directed by the Project Manager.

- Stone Cladding
- Flooring Works
- Dado Works
- Doors
- Paints
- Wooden Flooring/Sports Flooring/Vinyl Floor
- Fabric
- Built-In Furniture's
- Façade hard finish (Like Jali, Cladding etc)
- Other materials as directed by Project Manager/Architect.

Contractor shall follow the pour card/check list for all the concrete/finishing items on prescribed formats.

1.0   WOOD WORKS:

♦ **The rates for all items under this section include :**

1. All timber shall be of Ist class quality as described and indicated on drawings and schedules, shall be uniform in texture, free from large loose heads or cluster knots, injurious open shakes, bore holes, soft or spongy spots, hollow pockets and all other defects and blemishes. The size shown or described are to be taken as net sizes when finished, and cubic contents of wood as actually fixed would be measured.

2. All framing and other concealed wood members shall be sound wood or approved specials and shall be well seasoned. Wood more than 12% moisture content shall not be used in any part of the structure.

3. All joints shall be approved by Project mangers and according to the best practice. All joinery work shall be glued with best quality synthetic waterproof adhesive Fevicol or equivalent make as approved.

4. All woodwork in contact with masonry/ R.C.C shall be painted with approved antitermite paint before placing in position. Care shall be taken to keep the exposed surface clear. All concealed wood members shall be painted with applying Anti-termite paint.

5. Fixing the frames by using holdfasts, anchor fasteners, screws, nut and bolts etc.

6. The work shall have to be executed at all heights and levels.

7. Compliance with requirements of technical specification.

8. **2.0 FLOORING & CLADDING WORKS:**

   (The rates for all items under this section include) :

1. Use and waste of all temporary fillets, side forms, templates, moulds, straight edges etc.

2. Final preparation of the base, sub-grade or sub-floor including trimming of the base to remove all undulations including chipping or filling with concrete/ cement mortar, if necessary.

3. Cleaning and watering the surface immediately before laying the floor.

4. Providing bedding layer of mortar as specified, in the case of slabs, tiles etc. to correct levels and slopes as called for.

5. Cutting, Rubbing and Tinoxide (mirror) polishing. Rounding of corners, edges, and junctions of floor with skirting or dado.

6. Necessary cutting for Electrical switch boxes/ lift boxes, sinks, mixers, taps, gratings/floor trap jali etc and all other Electrical and Plumbing fixtures as deemed necessary by Project Manager.

7. Providing and applying silicon sealant between the joints of floor & dado, around bath tub, kitchen sink, wash basin etc.

8. Forming rounded recess in the floor where called for.

9. Providing grooves chamfering, moulding, noising, edge polishing etc where shown in the drawings.

10. Work in narrow widths, bands, architraves, curved surfaces, set backs, offsets, (any geometric pattern) pattern and design at all heights and locations, unless otherwise mentioned.

11. Curing, protecting and cleaning all surfaces as specified.

12. Leaving or chasing recess for skirting to match with plaster surface, bends, steps etc. on walls/ concrete and making them good after finishing of the wall.

13. Payment shall be made only for the finished dimension actually measured at site of work.

14. Matching grains/veins for different finishing items.

15. Protection (in required manner) & precaution to avoid any damage of flooring, dado, risers, treads etc. Any damage shall be repaired at contractors expenses.

16. Cost of pre polishing (Tinoxide)/mirror wherever pre polished stones are specified.

17. No work shall be started until the concealed piping drains etc, are laid by other agencies. Prior to commencing any finishing surfaces, levels and the slopes to drains shall be got approved, in writing by the Project Manger. Further to do so may result in demolishing the finished surface and redoing the work, all at contractors expenses

18. Laying floors to required slope in any size and shape of panels, pattern & design of panels. The strips shall not be paid for separately.

19. Fixing of clamp shall be with fastner (as approved by Project Manager) in case of RCC and in Brick Work pocket (100mmX75mmX75mm) shall be made and filled by non-shrinkage compound (as approved by Project Manager). The size and design of the cramp and fastener shall be to suite site requirement and shall be approved by the Project Manager.

20. Work shall be carried out at any elevations all heights, levels, leads and lifts.

21. Use of all scaffolding and cradles, dust seats and other coverings for fittings, furniture, floors etc. (for all heights and locations).

22. Painting and Polishing on stone/wood/structural steel prior to handing over.

23. Cladding in toilets and kitchen to be as per the pattern shown in drawings. To facilitate the work and its accuracy providing tile module and marking of location coordinates for electrical and plumbing fixtures in each toilet & kitchen.

24. The flooring stone / tile shall be taken continuously over the floor traps and cleanout plugs the location of which shall be clearly marked / engraved on the floor. The cut outs shall be made for fixing jali after completion the floor polish.

25.

26. Necessary cutting for Electrical switch boxes/ lift boxes, sinks, mixers, taps, gratings/floor trap jali , Railing Balusters, making holes for railing etc and all other Electrical and Plumbing fixtures as deemed necessary by Project Manager.

27. Work in narrow widths, bands, architraves, curved surfaces, set backs, offsets, (any geometric pattern) pattern and design at all heights and locations, unless otherwise mentioned.

28. Protection (in required manner) & precaution to avoid any damage of flooring, dado, risers, treads etc. Any damage shall be repaired at contractors expenses.

29. Unloading and staking of tiles at site, by manual means. Rate includes all labour charges, tools, plant, lead, lift and required housekeeping for stacking of material.

30. Quoted rate included all type of wastage including design, pattern etc.

31. Necessary cutting for Electrical switch boxes/ lift boxes, sinks, mixers, taps, gratings/floor trap jali etc and all other Electrical and Plumbing fixtures as deemed necessary by Project Manager.

32. Compliance with requirements of technical specification.

3.0 FINISHING (WALL & CEILING) WORKS & PLASTER WORKS:

♦ **The rates for all items under this section include :**

1. Making all construction and expansion joints, curving, curing.

2. Making grooves of any pattern as per drawings or as directed by the Project Manger in plaster and dados including rounding of junctions with floors.

3. In case of wall plaster, dado and skirting, raking out joints, cleaning the surfaces, application of cement slurry, applying plaster, skirting and dado treatment unless otherwise specified.

4. Work on patches, narrow widths, small quantities, curved surfaces, projected/ resets bands, setbacks, offsets, corbels, architraves etc.

5. Extra thickness of plaster over indentations etc.

6. Repairing and finishing the junctions of skirting and dado, with relevant mortar/finish.

7. Finishing the chases, edges of electrical fittings and boxes etc.

8. Use of all scaffolding and cradles, dustsheets and other coverings for the protection of fixtures, fittings, furniture, floors etc. (for all heights and locations).

9. Grooves, bands in plaster, on RCC bands, drip coarse etc. in plaster works as per directions.

10. Cleaning of paint splashes, drops or dirt, glasses, joinery, electric fittings etc., including washing the floor and leaving the premises neat and clean.

11. Work shall be carried out at any elevations all heights, levels, leads, lifts.

12. Enamel paint/ melamine polish application by compressed spray method. All other paint shall be finished with roller.

13. Cutting line of two different finishes should be in straight line or as shown in drawing, bands wherever required.

14. Providing and fixing 300mm wide chicken wire mesh with GI screw and washers at the junctions of two different materials and on all chasing for electrical & plumbing conduits, pipes, AC copper pipies etc.

15. Cost of waterproofing compound wherever mentioned in Bill of Quantities.

16. Extra thickness of plaster over indentations etc.

17. Finishing the chases, edges of electrical fittings and boxes etc.

18. Preparation and applying jamb wall & ceiling plaster surface.

19. Preparation, making & repair of opening holes of max. area 0.5 Sqm.

20. Use of all scaffolding and cradles, dustsheets and other coverings for the protection of fixtures, fittings, furniture, floors etc. (for all heights and locations).

21. Preparation & making of jamb plaster for Pre-fabricated Doors & windows as directed.

22. Preparation, Closing, making opening in Kitchen as per site requirement for AC and other services works.

23. Any repair work in all type of plaster balance works comprising with bonding agent etc complete as directed by Project Manager.

24. Compliance with requirements of technical specification.

4.0 MISCELLANEOUS WORKS:

♦ **The rates for all items under this section include :**

1. Steel forging, reducing to required shape, size and figure, drilling, tapping, counter sinking for screws, filling etc. and satisfactory workmanship required to fabricate, finish, erect and fix in position, all structural steel and iron in a good and perfect manner.

2. Neatly cut and fitted notches, true and squared ends including planings to sheared end of plates, angles etc., as directed.

3. All necessary templates, pattern moulds, mock-ups etc.

4. All welding/bolting/riveting as required and as shown, but weight of weld not to be paid.

5. Necessary hoisting equipment, temporary supports, scaffolds, bracings, connections, required for fabrication and erection including removing same after fixing local work in final position as per drawings.

6. Preparation of surface de-rusting, phosphating and application of primer coat.

7. Providing all bolts and nuts including holding down and anchor bolts, round, squared or tapered washers, anchor plates, rivets, packing pieces, gusset

plates, cleats, wedges, brackets, separators etc. (net weight to be computed and paid).

8. Providing all spikes, nails, service bolts, clamps, jigs etc.

9. Making all necessary templates, patterns moulds and platforms for layout etc.

10. All smithy work, unloading, getting in, hoisting, erecting and fixing in position at all heights, levels and locations, curve portion.

11. Rigidly inserting and setting in lead or other specified material and fixing into concrete and / or building into brick work while the work proceeds and for all fixing, anchoring, plugging, screwing, bolting etc. including non shrink grout and sealants as may be required or directed.

12. Painting two top coat of Red oxide primer before hoisting and erecting in position.

13. The priming coat is required to be of high grade loosing approved quality of Red Oxide primer to provide a coating having a good rust preventive properties and shall adhere well to the metal surfaces, affording a good foundation for subsequent coats.

14. Bending to required shape of square bars, pipes, angles, plates etc as per drawing.

15. Rigidly inserting and setting in lead or other specified material and fixing into concrete and/or building into brick work while the work proceeds and for all fixing, anchoring, plugging, screwing bolting etc. including non shrink grout and sealants as may be required or directed.

16. One shop coat of primer and protective treatment as specified.

17. Compliance with requirements of Technical Specifications.

18. Weight of packing material such as lead, quantity of cement grout etc. will not be measured.

19. Manufacture of railings thru welded or mechanical connections, as directed by the Architect

20. Compliance with requirements of technical specification.

LIST OF APPROVED MAKES OF MATERIALS

1. Cement (OPC) -43 grade - Ambuja, Birla, Ultratech,
    L & T, Lafarge

2. White Cement -   Birla, J.K

3. White washing lime - Dehradun (Source)

4. Paints   -   Asian Paints, ICI, Nerolac

5. Water proof cement paint -   Snowcem India Ltd

6. Fire Retardant paint -   Viper   or   approved
    equivalent

7. Epoxy - Fosroc/ STP/ Cico.

8. Glass - Modi Guard, Asahi or approved equivalent

9. Mirror - Modi Guard or approved equivalent

10. Waterproof Ply - Archid, Durain, Merino, Century

11. Commercial ply - Archid, Durain, Merino, Century

12.   Veneer - Archid, Durain, Merino, Century

13.   Laminate - Archid , Durain   Merino, Century or approved equivalent

14.   Cement Bonded Board - 'BISON' by NCL or approved equivalent

15.   Gypsum board. - India Gypsum Ltd.

16.   False Ceiling Members - Gyp. Steel of India Gypsum Ltd.

a) Perimeter
b) Celling section
c) Intermediate
d) Angle

17.   Flush Door Shutter   -   Archid, Durain, Century, Merino

18.   Door Hardware   -   Dorset/Haffle / Hetitch

19.   Door Lock & Handle   -
                                            Dorset/ Haffle   / Hetitch

20.	Door closer	-	Dorset/Haffle / Hetitch
21.	PVC strips	-	Fixopan or approved equivalent
22.	MDF	-	NUWUD MDF grade I as per IS 12406 / Green Panelmax/Bajaj Ecotec
23.	Screws, Nails etc.	-	Nettlefold or approved equivalent.
24.	Shower Cubical Hardware's or Approved equivalent	-	Dorma

METHODOLOGY/SPECIFICATIONS FOR WATERPROOFING

1.WATERPROOFING FOR BASEMENT PCC APP MEMBRENE WATERPROOFING BY HEATING PROCESS.

➢ The surface/wall shall be cleaned and all dust, foreign matters, loose materials or any other deposits which causes contamination will be removed.

➢ Treating the corners and junction of Floor and Wall area by making side coving in CM 1:4 admixed with FOSROC CONPLAST WL.

➢ If Any Major cracks / Construction joint is found, the same should be opened (10mm) and Filled with Repair mortar or with fixing injection grouting point along the construction joint with an interval of 1.5M c/c and grouting the same with cement slurry admixed with waterproofing compound.

➢ If any leakage is found then fixing injection grouting point at leakage areas and grouting with cement slurry admixed with waterproofing compound FOSROC CEBEX 100.

➤ Priming: Apply one coat of Bituminous primer on cleaned surface/wall

➤ Unroll the app membrane (4mm thick) Roll after application of primer.

➤ Align the membrane roll correctly & re-roll it in alignment before torching.

➤ Use gas burner to heat substrate & underside to softening points. When the embossing disappears, roll forward & press firmly against substrate to bond from the lower end towards the higher end.

➤ Keep overlap margin for minimum 100mm.

➤ Heat both the overlaps & use round tipped trowel to seal overlap. Excess compound should be smoothened & pressed into seam using hot trowel.
➤ Same process for side/vertical wall also applicable.

2.BASEMENT WATERPROOFING (BOX TYPE).
❖ Providing and laying of external waterproofing treatment for the basement, integral type cement based as per specification generally consisting of the following.

FLOOR: Laying waterproofing treatment on the PCC floor to manufacturers specification and covering the same with Kadapa slabs of 20 mm thick with SINGLE LAYER.

- ➢ Necessary waterproof pointing in the joints, keeping pressure release pipes at suitable intervals to keep the waterproofing layer in position if found necessary. (Admix with FOSROC CONPLAST 421IC)

- ➢ Providing waterproof treatment on wall, as the manufacturers specification covered with Kadapa slab 20mm thick.

Finally, 20mm thick waterproofing plaster will be done on floor and side wall admixed with waterproofing compound CONPLAST 421 IC.

Thanks for Reading